T0224253

Embracing Risk:
Cyber Insurance
as an Incentive Mechanism for
Cybersecurity

Synthesis Lectures on Learning, Networks, and Algorithms

Editor
Lei Ying, *University of Michigan, Ann Arbor*

Editor Emeritus
R. Srikant, *University of Illinois at Urbana-Champaign*

Founding Editor Emeritus
Jean Walrand, *University of California, Berkeley*

Synthesis Lectures on Learning, Networks, and Algorithms is an ongoing series of 75- to 150-page publications on topics on the design, analysis, and management of complex networked systems using tools from control, communications, learning, optimization, and stochastic analysis. Each lecture is a self-contained presentation of one topic by a leading expert. The topics include learning, networks, and algorithms, and cover a broad spectrum of applications to networked systems including communication networks, data-center networks, social, and transportation networks. The series is designed to:

- Provide the best available presentations of important aspects of complex networked systems.

- Help engineers and advanced students keep up with recent developments in a rapidly evolving field of science and technology.

- Facilitate the development of courses in this field.

Embracing Risk: Cyber Insurance as an Incentive Mechanism for Cybersecurity
Mingyan Liu
2021

Edge Intelligence in the Making: Optimization, Deep Learning, and Applications
Sen Lin, Zhi Zhou, Zhaofeng Zhang, Xu Chen, and Junshan Zhang
2020

Poisson Line Cox Process: Foundations and Applications to Vehicular Networks
Harpreet S. Dhillon and Vishnu Vardhan Chetlur
2020

iv

Age of Information: A New Metric for Information Freshness
Yin Sun, Igor Kadota, Rajat Talak, and Eytan Modiano
2019

Multi-Armed Bandits: Theory and Applications to Online Learning in Networks
Qing Zhao
2019

Diffusion Source Localization in Large Networks
Lei Ying and Kai Zhu
2018

Communications Networks: A Concise Introduction, Second Edition
Jean Walrand and Shyam Parekh
2017

BATS Codes: Theory and Practice
Shenghao Yang and Raymond W. Yeung
2017

Analytical Methods for Network Congestion Control
Steven H. Low
2017

Advances in Multi-Channel Resource Allocation: Throughput, Delay, and Complexity
Bo Ji, Xiaojun Lin, and Ness B. Shroff
2016

A Primer on Physical-Layer Network Coding
Soung Chang Liew, Lu Lu, and Shengli Zhang
2015

Sharing Network Resources
Abhay Parekh and Jean Walrand
2014

Wireless Network Pricing
Jianwei Huang and Lin Gao
2013

Performance Modeling, Stochastic Networks, and Statistical Multiplexing, Second Edition
Ravi R. Mazumdar
2013

Packets with Deadlines: A Framework for Real-Time Wireless Networks
I-Hong Hou and P.R. Kumar
2013

Energy-Efficient Scheduling under Delay Constraints for Wireless Networks
Randall Berry, Eytan Modiano, and Murtaza Zafer
2012

NS Simulator for Beginners
Eitan Altman and Tania Jiménez
2012

Network Games: Theory, Models, and Dynamics
Ishai Menache and Asuman Ozdaglar
2011

An Introduction to Models of Online Peer-to-Peer Social Networking
George Kesidis
2010

Stochastic Network Optimization with Application to Communication and Queueing
Systems
Michael J. Neely
2010

Scheduling and Congestion Control for Wireless and Processing Networks
Libin Jiang and Jean Walrand
2010

Performance Modeling of Communication Networks with Markov Chains
Jeonghoon Mo
2010

Communication Networks: A Concise Introduction
Jean Walrand and Shyam Parekh
2010

Path Problems in Networks
John S. Baras and George Theodorakopoulos
2010

Performance Modeling, Loss Networks, and Statistical Multiplexing
Ravi R. Mazumdar
2009

Network Simulation
Richard M. Fujimoto, Kalyan S. Perumalla, and George F. Riley
2006

Embracing Risk: Cyber Insurance as an Incentive Mechanism for Cybersecurity

Mingyan Liu

ISBN: 978-3-031-01253-2   paperback
ISBN: 978-3-031-02381-1   ebook
ISBN: 978-3-031-00245-8   hardcover

DOI 10.1007/978-3-031-02381-1

A Publication in the Springer Nature series
SYNTHESIS LECTURES ON ADVANCES IN AUTOMOTIVE TECHNOLOGY

Lecture #26
Editor: Lei Ying, *University of Michigan, Ann Arbor*
Editor Emeritus: R. Srikant, *University of Illinois at Urbana–Champaign*
Founding Editor Emeritus: Jean Walrand, *University of California, Berkeley*
Series ISSN
Print 2690-4306   Electronic 2690-4314

Embracing Risk: Cyber Insurance as an Incentive Mechanism for Cybersecurity

Mingyan Liu
University of Michigan

SYNTHESIS LECTURES ON LEARNING, NETWORKS, AND ALGORITHMS
#26

ABSTRACT

This book provides an introduction to the theory and practice of cyber insurance. Insurance as an economic instrument designed for risk management through risk spreading has existed for centuries. Cyber insurance is one of the newest sub-categories of this old instrument. It emerged in the 1990s in response to an increasing impact that information security started to have on business operations. For much of its existence, the practice of cyber insurance has been on how to obtain accurate actuarial information to inform specifics of a cyber insurance contract. As the cybersecurity threat landscape continues to bring about novel forms of attacks and losses, ransomware insurance being the latest example, the insurance practice is also evolving in terms of what types of losses are covered, what are excluded, and how cyber insurance intersects with traditional casualty and property insurance. The central focus, however, has continued to be risk management through risk transfer, the key functionality of insurance.

The goal of this book is to shift the focus from this conventional view of using insurance as primarily a risk management mechanism to one of risk control and reduction by looking for ways to re-align the incentives. On this front we have encouraging results that suggest the validity of using insurance as an effective economic and incentive tool to control cyber risk. This book is intended for someone interested in obtaining a quantitative understanding of cyber insurance and how innovation is possible around this centuries-old financial instrument.

KEYWORDS

cyber insurance, cyber risk quantification, contract theory, incentives, game theory

Contents

Preface xiii

Acknowledgments xv

1   Introduction: What is Insurance and What is Cyber Insurance? 1
   1.1   What is an Insurance Contract? 2
      1.1.1   Risk Neutrality and Risk Aversion 3
      1.1.2   Moral Hazard, Adverse Selection, and Premium Discrimination 5
   1.2   What is Unique about Cyber Insurance? 6
   1.3   Security as a Public Good and Insurance as Risk Control 9
   1.4   Voluntary Participation 10
   1.5   Related Literature 10
   1.6   Roadmap 11
   1.7   Intended Audience 12

2   A Basic Cyber Insurance Contract Model 15
   2.1   A Single-Agent, Single-Period Model 15
      2.1.1   The Model 15
      2.1.2   Risk Neutrality vs. Risk Aversion 19
      2.1.3   The Role of Pre-Screening 22
      2.1.4   Summary 23
   2.2   A Single-Agent, Multi-Period Model 24
      2.2.1   A Two-Period Contract Model 24
      2.2.2   Two-Period Contract Design with Post-Screening 24
      2.2.3   Two-Period Contract Design with Pre-Screening 26
      2.2.4   State of Security and Optimal Contracts 27
   2.3   Numerical Results 27
      2.3.1   Impact of Agent's Risk Attitude γ 27
      2.3.2   Impact of Pre-Screening Noise 29
      2.3.3   Exponential Loss and Uniform Pre-Screening Noise 29
   2.4   Chapter Summary 29

2.5   Table of Notations Used in this Chapter 31
2.6   Appendix 31
     2.6.1   Proof of Theorem 2.3 31

3   Insuring Clients with Dependent Risks **37**
   3.1   A Model of Two Agents 38
   3.2   Two Risk-Neutral Agents 39
     3.2.1   Case (i): Neither Agent Enters a Contract 39
     3.2.2   Case (ii): One and Only One Enters a Contract 41
     3.2.3   Case (iii): Both Agents Purchase a Contract 42
     3.2.4   Optimal Contracts for Two Risk-Neutral Agents 44
   3.3   Two Risk-Averse Agents 46
   3.4   Multiple Agents, Correlated Losses, and the Insurer's Risk Aversion 47
   3.5   Numerical Results 48
     3.5.1   Impact of Pre-Screening Noise 48
     3.5.2   The Sufficient Conditions of Theorem 3.2 50
     3.5.3   Loss and Pre-Screening Noise Distributions 52
     3.5.4   Correlated Pre-Screening Noises 53
   3.6   Chapter Summary 54
   3.7   Tables of Notations Used in This Chapter 55
   3.8   Appendix 57
     3.8.1   Finding the Solution to the Best-Response Function $B_i^{oo}(e_{-i})$ 57
     3.8.2   Finding a Fixed Point 58
     3.8.3   Proof of Theorem 3.1 Parts (ii) and (iii) 58

4   A Practical Underwriting Process **63**
   4.1   Computing Premiums Using Base Rates 64
   4.2   The Insurance Policy Model and Analysis 68
     4.2.1   Base Premium Calculation 68
     4.2.2   The Security Incentive Modifier 69
     4.2.3   Mapping Security Incentive to Probability of Loss 70
     4.2.4   The Insurer's Utility/Profit Function 70
     4.2.5   Analysis of the Optimal Incentives and Carrier Utility/Profit 71
     4.2.6   Portfolio C and Third-Party Liability Clauses 73
     4.2.7   Summary of Our Findings 76
   4.3   Numerical Examples 76
     4.3.1   Examples of the Loss Probability Function 77

	4.3.2	Examples of the Loss Distribution	78
	4.3.3	Example 1: An SP and a Customer with Large Revenue	79
	4.3.4	Example 2: An SP and Multiple Customers with Smaller Revenue	81
4.4	Discussions		85
	4.4.1	Is the Premium Discount Sufficient?	85
	4.4.2	Social Welfare	88
	4.4.3	Modeling Third-Party Liability	90
	4.4.4	Non-Monopolistic Insurer	91
4.5	Chapter Summary		91
4.6	Table of Notations Used in This Chapter		91
4.7	Appendix		93
	4.7.1	Proof of Theorem 4.3	93

5	**How to Pre-Screen: Risk Assessment Using Data Analytics**	**95**
5.1	Predictive Power of Measurement Data	96
5.2	Cyber Incident Forecast	101
5.3	Fine-Grained Prediction	105
5.4	Bringing Technology to Market	108
5.5	Chapter Summary	111

| 6 | **Open Problems and Closing Thoughts** | **113** |

| | **Bibliography** | **117** |

| | **Author's Biography** | **127** |

Preface

The oldest form of insurance is generally thought to be marine insurance, and a written example of such is found in the Code of King Hammurabi carved into a Babylonian monument dated 18th century B.C., which stipulates that "If a merchant receives a loan to fund his shipment, he would pay the lender some money in compensation for the lender providing a guarantee that he would cancel the loan if the shipment sank or was stolen." [45] This book is on the subject of cyber insurance, one of the newest forms of insurance created in response to increasing cyber threats and increasing impact that information security has on business operations.

ROADMAP

After a short introduction to insurance in general and cyber insurance in particular (Chapter 1), we will start with a simple mathematical model describing the relationship between an insurer and an insured, and how the contract is determined between them (Chapter 2). We will use this model to illustrate the importance of quantitative risk assessment, the difference it makes in determining the pricing of an insurance policy, and the impact it has on the cyber risk of the insured. We then take a deeper dive into this contract framework and explain why risk dependence, while certainly complicating matters, also emerges as an opportunity that can and should be exploited to the mutual benefit of both the insurer and the insured (Chapter 3). How our analysis and insights translate into actual underwriting practices is also described using a very common underwriting workflow process (Chapter 4).

To compliment this discussion on contract design, which relies on our ability to assess cyber risks at a firm level, we also outline a number of ways to quantify cyber risks in a global and scalable manner (Chapter 5) and the ensuing commercialization effort in this space and how our data analytics tools are being used today.

The book ends with a number of open challenges in this domain (Chapter 6).

NOTE ON STYLE AND READABILITY

We will use mathematical equations to describe models—we need the precision upon which to base our discussion. We will, however, largely avoid lengthy mathematical derivation or presenting formal, technical results, and opt for intuitive and qualitative explanations of these results instead. We provide all necessary references to the technical papers for an interested reader.

There are many parties to an insurance policy: the enterprise/insured purchasing it, the underwriter selling it, the broker trying to match a buyer and a seller, the reinsurer providing insurance for the underwriter, and additional parties such as auditors and providers of risk and forensic analysis, among others. This book takes an outsider's holistic view of the most important part of this eco-system, i.e., the insured and the insurer and how their relationship impacts the state of cybersecurity for all. For this reason, we almost always consider the welfare of both parties, as well as a more objective measure of security. For simplicity of presentation, we will ignore the existence of the broker and roll into one the roles of insurer and reinsurer.

INTENDED AUDIENCE

This book is intended for someone interested in obtaining a quantitative understanding of cyber insurance and how innovation is possible around this centuries-old financial instrument. They broadly include the following.

1. Researchers working in the areas of cybersecurity, cyber risk quantification, and the economics of information security: they will be able to get a much better understanding of how cyber risk is quantified and managed in the context of insurance, and in particular, Chapters 4 and 5 will give them a very concrete sense of how it is practiced in reality. Such foundational knowledge should inspire ideas and open up new avenues of pursuit in their own research domains.

2. Researchers working in the area of mathematical economics and, more specifically, incentive mechanisms: they may be interested in learning how insurance as a type of economic lever is used in the context of cybersecurity and its unique challenges, especially from Chapters 3 and 4.

3. Practitioners of cyber insurance on all sides, underwriters, brokers, reinsurers, and corporate risk managers who purchase insurance: some of them may not have a cybersecurity background, in which case they should find this book useful in providing some domain knowledge; some of them may be well-versed in both cybersecurity and (traditional) insurance products, in which case I hope the main premise of the book, that insurance could be used more effectively as a risk reducing mechanism, will give them fresh food for thought, and perhaps motivate them to adopt this line of thinking and seek similar innovation in their respective practices.

Some basic probability background would certainly help a reader get the most out of this book (risk after all is all about probability) but is by no means necessary. All technical results are accompanied by copious amount of explanation, so one could choose to skip the mathematical expositions and go straight for the plain language.

Mingyan Liu
April 2021

Acknowledgments

Much of the research presented in this book started as a project funded by the Department of Homeland Security (DHS). The three-year project was entitled "Towards a Global Network Reputation System: A Mechanism Design Approach" and started in 2013. The basic idea was to provide an org-level quantitative cybersecurity posture assessment framework using the notion of "reputation", by developing tools for Internet measurement and advanced data analytics.

Here I would like to sincerely thank my program manager, Dr. Doug Maughan, who really took a chance with our initial concept. I remember talking to him on the phone in early 2013, trying to convince him of the merit of the idea. He saw the promise but also the risk in its direction—it was an unusual approach—"Are you sure this reputation thing is the right way to go?" I reassured him, with as much confidence as I could muster, but who could know for certain it was going to work? It did take us another two years before we could get our paper accepted. Doug took a risk on an unconventional idea, and it paid off. This initial concept evolved and developed into an enterprise cybersecurity ratings system, and this project led to the eventual technology transition and commercialization success. None of this would have been possible without this project and the funding support from the DHS.

This initial project also led to our work on policy design: how to utilize firm-level risk quantification to design more proactive and preventative measures, and to design mechanisms to incentivize better behavior, more investment in security, and adoption of best practices. This became a big part of my research activities and constitutes the majority of this book. I am grateful for the support I have received from both the DHS and the National Science Foundation (NSF) for this work.

I would not have been able to accomplish any of the research presented here without a group of fantastic collaborators. My gratitude goes to my co-authors on papers that form the foundation of the body of research contained in this book. They include the following former and current Ph.D. students of mine, in the order of their graduation date:

- Yang Liu (now a professor at UC Santa Cruz),

- Parinaz Naghizadeh (now a professor at the Ohio State Univ.),

- Armin Sarabi (now a research scientist at the Univ. of Michigan),

- Mohammad Mahdi Khalili (now a professor at the Univ. of Delaware),

- Xueru Zhang (a Ph.D. student at the Univ. of Michigan).

My gratitude also goes to two key collaborators: Professor Michael Bailey of the Univ. of Illinois, Urbana-Champaign, and Dr. Sasha Romanosky of the Rand Corporation.

This acknowledgment cannot be complete without mentioning my husband of 20 years, Manish Karir, who is both a co-author, co-inventor, and the co-founder of QuadMetrics, the startup we built on the DHS funded research.

Mingyan Liu
April 2021

CHAPTER 1

Introduction: What is Insurance and What is Cyber Insurance?

Insurance as an economic instrument designed for risk management through risk spreading has existed for centuries. The oldest form of insurance is generally thought to be marine insurance, and a written example of such is found in the *Code of King Hammurabi* carved into a Babylonian monument dated 18th century B.C., which stipulates that "If a merchant receives a loan to fund his shipment, he would pay the lender some money in compensation for the lender providing a guarantee that he would cancel the loan if the shipment sank or was stolen" [45].

Cyber insurance is one of the newest sub-categories of this old instrument. It emerged in the 1990s in response to an increasing impact that information security started to have on business operations. While the idea seemed promising, the market failed to materialize in a substantial way, and these products remained a niche for unusual demands. Insurance coverage during most of these initial years was fairly limited, and clients were primarily small and medium businesses (SMBs) in need of insurance to qualify for tenders.

One of the first well-known cyber-liability policies was developed for the Lloyd's of London in 2000 [6, 39], which provided third-party as well as business interruption coverages. Other early entrants to the cyber insurance market included American International Group (AIG) and Chubb. At that time, it was thought that a major cyber risk would take the form of one firm inadvertently transmitting virus to and infect another firm, who would then bring suit against the first firm, as well as the first firm's own business interruption loss. This cyber-liability policy was one of the first to include both parties' coverages in the same policy.

Suits against organizations on this basis have proven to be rare, and the focus of policies that have been developed since 2000 has primarily been on business interruption (suffered by oneself or suffered by one's customers with appropriate service level agreements), payment of fines and penalties, credit monitoring costs, crisis management and public relations costs, and the cost of restoring or rebuilding private data, and now increasingly, ransom payment given the prevalence of ransomware attacks.

These policies continue to expand and evolve today and the market for cyber-insurance products has been growing steadily over the past 20 years [2, 3, 14, 46, 57, 61], with over 80 carriers around the world and total premiums on track to reach $10B by 2020 and approaching $15B in another few years. Typical premiums are estimated to start from $10K–$25K and go as high as $50M [33, 99]. These contracts are reported to have an average of $16.8M limit [86], with some coverage limits up to $200M–$300M [99]. Coverage on these policies typically include:

- direct first-party costs, e.g., forensic investigations;

- liability and third-party claims, e.g., legal defense; and

- various add-ons such as business interruption costs, compensation against social engineering attacks, etc.

These products enable organizations and businesses to manage their cyber risks by transferring (part of) their risks to an insurer in return for paying a premium as we detail below.

1.1   WHAT IS AN INSURANCE CONTRACT?

A contract is an agreement between two or more parties. In the context of this book we will limit ourselves to two parties: a *principal* and an *agent*. Mathematically, the interaction between these two parties is referred to as a *principal-agent problem* [43, 78],[1] which has been widely studied in the fields of contract theory, game theory and multi-agent systems [38, 50, 53, 91, 111]. The two parties in such a setting are also referred to as the first mover (or the leader) and the second mover (or the follower), respectively. The former dictates the format (terms) of a contract, some of which may be be determined based on the latter's actions/choices; the latter then takes an action which substantiates the contract terms that both sides agree on.

Some of the most typical of contracts we encounter in our daily lives have to do with service agreements: payment in exchange of service performed and goods delivered, such as domestic services, home improvement, food delivery, etc. A very simple example in such a setting is where the contractor offers a list of prices (and payment schedule) for different levels of services; the customer selects the desired service, thereby agreeing to the price to be paid and the associated payment schedule.

This is no different in the context of an insurance contract: the insurer (or the underwriter) is the first mover and the insured (or the agent) the second. The former specifies what a contract looks like and how its parameters are going to be determined (e.g., how the premium might change given the model and year of the vehicle to be insured, among other factors); the latter takes actions (e.g., selecting a particular model and year, among other factors),[2] which then determines the terms of the contract.

But whereas other contracts transact in tangible goods and services, insurance contracts transact in "risk," or more precisely, our natural desire to avoid risk and minimize uncertainty. An insured in this case enters into a contract (pay a premium) in order for the insurer to take

---

[1]It is alternately referred to as a leader-follower problem or a Stackelberg game [123].

[2]While we may not always think of insurance when purchasing a particular model and year, our choice in the latter does determine the former.

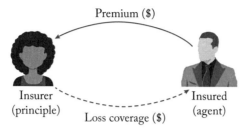

Figure 1.1: Two sides of an insurance contract.

on some or all of the insured's risk should an adverse event occur. More precisely, the contract spells out how the two are going to split the expenses incurred by the insured when specified events materialize. This is illustrated in Figure 1.1, where a solid curve denotes a certainty event (paying a premium) and a dashed curve denotes a random event (incurring a loss), a convention we will repeatedly use in diagrams throughout the book.

1.1.1   RISK NEUTRALITY AND RISK AVERSION

The very existence of an insurance market hinges on the notion of *risk aversion*.[3] The following simple example explains what risk aversion is in the context of insurance. Suppose we (as an individual or a firm) face an uncertain outcome, whereby with probability 0.5 nothing happens but with probability 0.5 we stand to lose $100. Also suppose we have at least $100 in the bank (so we are not worried about liquidity). Our expected loss is $50, the average between the two outcomes. If someone offers an insurance to fully protect against the second, adverse outcome, with a price (called premium) of $50, then in terms of *expectation*, we should be completely indifferent: with insurance, we stand to lose $50 with certainty; without insurance we stand to lose $50 in expectation. This indifference is referred to as being risk neutral.

However, we often find ourselves drawn to the *certainty* offered by insurance: yes, we pay out $50 for sure, but we will never have to pay anything above that! This preference for certainty, even though the two are exactly equivalent in terms of expectation, is referred to as being risk averse. More precisely, risk aversion dictates that we would actually be willing to pay a premium in an amount (say $60) exceeding $50 in exchange for avoiding uncertainty—whatever happens, we are looking at paying $60 and only $60. The risk is now entirely on the insurer: if the adverse event occurs, the insurer will cover the $100 loss (and thus suffer a net loss of $40); otherwise she pockets the entire $60. In expectation then, the insurer makes $10 in profit. This positive profit originates from our willingness to pay for someone else to take on our risk.

Intuitively, this suggests that we are placing a heavier emphasis on the risky (loss) event. The amount $60 is called the *certainty equivalent* (CE), i.e., this is the amount we are willing

---

[3]The notion opposite of risk aversion is *risk acceptance* or *risk seeking*. Whereas risk aversion is what drives people to purchase insurance, risk acceptance is what motivates people to gamble.

to lose with certainty in exchange for avoiding the uncertain loss. The difference between this certainty equivalent and the actual loss in expectation is called the *risk premium* ($10 in this example). The "premium" we pay for an insurance policy derives its name from the notion of risk premium, and is mathematically the certainty equivalent, or the expected loss plus the risk premium.

The opposite regime is characterized by a preference for uncertainty, i.e., unwillingness to purchase such an insurance at or above $50. For instance, maybe we are only willing to consider at the price $40. This attitude is referred to as *risk seeking* or *risk accepting*, which suggests that we are downplaying the risky (loss) event with respect to the expectation. Risk premium is similarly defined, but would be negative in this case. It should now be clear that from an insurer's point of view, in expectation, she would lose money on this $40 policy,[4] and that her business model is crucially sustained by the existence of risk-averse individuals and positive risk premiums. Since this book is entirely about insurance, we will no longer dwell on risk seeking.

The way we model risk aversion in our analysis is to adopt a concave, increasing utility function of one's assets (negative of one's loss/cost). A typical example is an exponential function $g(x) = -e^{-\gamma x}$ indexed by a parameter $\gamma > 0$ called the agent's risk attitude. This γ measures risk aversion: the higher γ, the more risk averse the agent. The agent's utility U is then a concave and decreasing function of his loss L, expressed as follows:

$$U(L) = g(-L) = -\exp\{-\gamma \cdot (-L)\}, \tag{1.1}$$

where L denotes the total amount of expenditure, a random variable, including the cost of purchasing insurance (premium payment), potential loss less payout from insurance, and a variety of other fixed costs (e.g., paying for firewall or other security measures).[5] This $U(L)$ is thus also a random variable and the agent's goal is to maximize its expectation, $E[U(L)]$, with respect to the distribution of L. In contrast, a risk-neutral agent would be represented by an affine utility of the exact same costs and expenditures:

$$U(L) = \kappa(-L), \tag{1.2}$$

for some parameter $\kappa > 0$, which we will often take to be $\kappa = 1$ without loss of generality.

Figure 1.2 depicts an example of the risk-averse utility in comparison with a risk-neutral utility, where the realized loss is between amounts L_1 and L_2 with some distribution. The risk-averse utility at the point $E[L]$ is $U(E[L])$, which is greater than the expected utility $E[U(L)]$ (this is the hallmark of a concave function), meaning this utility function associates higher value with certainty (if we know the loss is going to be $E[L]$) over uncertainty (the expected utility $E[U(L)]$ when L is random). The certainty equivalent, CE, is found by equating $E[U(L)] = U(CE)$, i.e., the fixed loss amount that would yield the same utility as the expected utility over

---

[4]This does not suggest that insurers themselves are risk neutral; indeed they are not. Their own risk aversion only makes it more important to seek individuals who are even more risk averse; this also drives them to seek reinsurance. See more discussion that follows.

[5]In later chapters we will break this down into different categories.

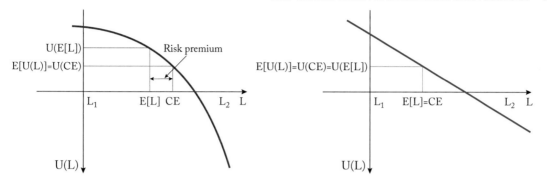

Figure 1.2: Difference between risk neutral (blue) and risk averse (red) utilities.

the uncertain loss. The difference is the risk premium the agent places on certainty (known loss CE) over uncertainty (L being anywhere between L_1 and L_2). By contrast, with a risk-neutral utility function, the expected utility is the same as the utility of the expected loss, resulting in zero risk premium.

As a general rule, we will assume the clients/agents are risk averse and the underwriters/principals risk neutral. The latter is in practice far from the truth—insurers are some of the most risk-averse folks in the world! They deal with this by seeking re-insurance to transfer the risks they have taken on from retail insurance to reinsurers. But more importantly, the concept of risk pooling gives them significant advantage in this business (mathematically, it means the losses they face become less uncertain and increasingly well-captured by the expectation). For that reason, we don't lose much, for the time being, by assuming that the other side of the contract is risk neutral. We will discuss risk accumulation and aggregation in Chapter 6—in the case of cyber, there is significant risk of correlated loss events that an insurer has to be aware of and guard against.

1.1.2 MORAL HAZARD, ADVERSE SELECTION, AND PREMIUM DISCRIMINATION

Cyber insurance inherits classic insurance problems of moral hazard (clients lower their investment in self-protection once they are insured) and adverse selection (higher-risk clients are more likely to seek protection and more likely to seek higher protection).

Moral Hazard refers to a lack of incentive to guard against risk when one is protected from its consequences, e.g., by insurance. This lack of incentive exists primarily because in general the insured's actions are unobservable to the insurer. If they were observable then the insurer could conceivably punish bad behavior (e.g., impose penalty if the insured fails to patch software vulnerabilities on time), thereby restoring the incentive for the insured to maintain its own risk protection. Indeed, the mere possibility of such observability could serve as a credible threat—if the insured knows its actions are observable by the insurer it might behave more responsibly.

Adverse selection refers to the situation that those in higher risk conditions are more likely to seek insurance for risk protection. Thus, in an unregulated market where not everyone is required by law to carry insurance, an insurer is more likely to see riskier clients. This issue has been repeatedly highlighted in the healthcare insurance discussion, that the insured pool skews heavily toward the sick and unhealthy if insurance is voluntary.

Both moral hazard and adverse selection fundamentally stem from an information asymmetry between the insured and the insurer: unknown effort and unknown type, respectively; in both cases the insured knows what the insurer does not. Our main focus in this book is on dealing with moral hazard and mitigate risk, and will assume that the insured's risk type is common knowledge.

One way to mitigate moral hazard is through the use of *premium discrimination*, whereby the insurer attempts to use the premium as a lever: lower it to reward better effort (or lower risk) on the insured's part, and increase it to punish lack of effort (or higher risk). The insurer can accomplish this through procedures such as an audit or penetration tests, and very often does so based on the insured's past history (whether it had breaches in the past and how many). How to do so in a scalable and efficient way is the subject of Chapter 5. When premium discrimination is done effectively, cyber insurance may be viewed as both a method for transferring/mitigating cybersecurity risks [58, 101, 115] and a potential incentive mechanism for internalizing the externalities of security investments.

1.2   WHAT IS UNIQUE ABOUT CYBER INSURANCE?

Both moral hazard and adverse selection are issues that exist in all insurance products, so while we do have to be mindful of them, it is worth pointing out a number of more distinct features of cyber insurance that we will need to pay attention to in our modeling effort.

Limited actuarial data and lack of domain knowledge to determine risk and liability.   Unlike other, more mature areas of insurance, such as life, health, travel, home, property, and auto, where the industry has accumulated decades if not centuries of data to enable sophisticated statistical modeling and risk analysis, there has been a dearth of data in cyber. Data breaches have not been widely reported for long, and neither their frequency nor their size has reached a seeming steady state. Their emergence as a common occurrence receiving wide media attention only started within the last 7–8 years. A prime example of an early cyber incident that received intense reporting was the Target data beach (late 2013), whose magnitude has by now been dwarfed by those of Sony, Equifax, and Marriott, among others. By contrast, even though we don't hear very much about the niche area of insurance for kidnapping for ransom, it is a centuries-old crime and underwriters specializing in ransom insurance have very good statistical models [110].

Indeed, a persistent lack of actuarial data has been the most critical obstacle in the wider adoption of cyber insurance and fully realizing its risk mitigation potential. The insurance industry often resorts to creative defensive measures that focus on self-protection of the carrier

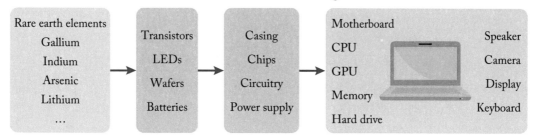

Figure 1.3: A partial supply chain in computer manufacturing.

itself, such as excessive exclusions/restrictions and expensive premiums on one hand, and very limited due diligence in customer surveys/audits on the other hand. On the demand side, the lack of standard risk assessment and management metrics and tools has led to widespread lack of confidence on the amount of protection one ultimately gets from insurance, and resulted in confusion on how to balance investing in classical security measures and in cyber insurance. This has in no small degree contributed to the fact that despite rapid growth, cyber insurance remains a nascent market.

Associated with the lack of actuarial data is a general lack of domain knowledge among underwriters when it comes to cyber risk assessment; this is to be expected—cyber risk is an emerging type of risk, and even the cybersecurity community itself is generally more concerned with security than with risk assessment.

Cyber risks are heavily interdependent.   In other insurance domains a conditional independence property is generally true: for instance, given two individuals driving vehicles of the same model/year and living in the same area with similar commute, their likelihoods of auto accident and damage are independent of each other. Therefore, we can treat them separately when determining their individual policies. The situation is much messier in the cyber insurance space. Large scales of outsourcing lead to complex supply chain and vendor relationships among entities. Figure 1.3 shows a typical and partial example of a supply chain, that in the manufacturing of a computer: many steps and many vendors are involved. In manufacturing, supply chain risk often refers to a scenario where the failure of one (perhaps very small) component on the chain leads to the failure of the ultimate product.

Similar outsourcing in services leads to similar supply chain risks and complex risk dependencies in the context of cyber insurance, whereby a firm's cyber risk is not just the result of its own actions, but that of the actions of its vendors/third parties and fourth parties. A prime example is a company using a service provider (e.g., AWS); if the latter is breached and goes down, the former suffers business interruption. This is illustrated in Figure 1.4, where firms outsource a variety of services such as accounting, payroll, email, cloud, storage, in addition to purchasing software products; these services in turn rely on their own service providers and so on. Indeed, many of the recent high-profile data breaches (e.g., the aforementioned Target breach

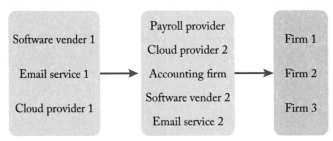

Figure 1.4: Cyber risk dependencies.

and the 2014 JP Morgan Chase data breach) are vendor induced breaches. This risk dependency introduces an interesting challenge: how do we price policies meant for individual companies, knowing their risks are correlated?

Correlated value-at-risk (VaR) [98].   Because of the risk dependency, when estimating the value at risk an insurer is exposed to, one must take into account the fact that many loss events will be correlated and cannot be treated as independent. An example of this is also shown in Figure 1.4, where some service providers simultaneously serve a large number of firms, effectively causing the latter's business interruption losses to be correlated.

Fast-changing cyber risk landscape compared to other risks.   In a more traditional domain like home, property, or auto, risk may be treated as a stationary process: we don't change cars or alter our driving habits on a daily basis; health risks of life style choices are reasonably well understood and relatively slow-changing.[6] Cyber risk is an entirely different beast in this regard, where breaches of different kinds and enormity appear as discrete events and highly unpredictable. Examples of such abound: data breaches set off by geopolitical events, orchestrated by cyber vigilantes, and copycat actions motivated by other major cyber attacks and so on. Vulnerabilities in our computing systems are also changing on a daily basis depending on what is disclosed, what has been patched, and what has an exploit kit in the wild. This makes it extremely difficult to build good statistical models for forward-looking, risk prediction purposes, on which insurance underwriting relies.

An additional observation.   Having laid out these challenges, it is worth noting that cyber risks, as well as risk controls, are both man-made unlike some other risk domains. This makes it all the more important to understand how such instrument should be designed. Herein lies an opportunity to use them as economic and incentive tools to induce better and more responsible behavior by corporate citizens.

---

[6]The forecasting of extreme weather in a warming climate may be an exception.

1.3 SECURITY AS A PUBLIC GOOD AND INSURANCE AS RISK CONTROL

It has been observed since the early 1990s that traditional vendor solutions to improve security were not going to be enough to ensure a desirable state of cyber-security [30]. One key reason behind this is the *public goods* nature of security investment. Cyber risk dependency means that efforts made by one entity not only serves to protect its assets against security threats, but also benefits other interacting entities (e.g., the former may be less likely to be infected and used as a launching pad of attacks against the latter, or the former may be less likely to cause business interruption to the latter).

In other words, one firm's expenditure or effort in security in an interconnected system provides *positive externalities* to others, much like investment in clean water, clean air, public libraries, etc. Consequently, the provision of security may be viewed as a problem of *public good* provision. Formally, a public good is defined as a non-rivalrous commodity—its use by one does not reduce its availability to others [87]. It is well known that in the absence of regulation, the provision of *public goods* is in general inefficient [87]. To eliminate this inefficiency, the literature on security investment involving rational decision makers has proposed regulating mechanisms for improving the level of cybersecurity to its socially desirable levels. Examples of such mechanisms include introducing subsidies and fines based on security investments [55, 75], assessing rebates and penalties based on security outcomes [55], and imposing a level of due care and establishing liability rules [75, 119].

Within this context, premium-discriminating cyber insurance contracts are viewed as one of such regulating mechanisms [66], and is perhaps the only free-market economic instrument for this purpose. However, insurance is fundamentally a *risk transfer* instrument—we pay a premium for someone else to take over all or part of our risk, and in that process we give up our own incentive to take positive actions, i.e., moral hazard—it is therefore imperative that we ask the question of whether there is more to risk transfer: can we turn insurance into a *risk reduction* instrument? Our last observation in the previous section points to the fact that, unlike risks associated with natural disasters, cyber risks are entirely induced by human activities (or inactivities). Therefore, it is appropriate to look for ways to reduce that risk.

Our goal then in this book, is to shift the focus from risk transfer to risk reduction, and look for ways to re-align the incentives so insurance can become an effective risk control mechanism. On that front we have encouraging results as presented in the chapters that follow. The content is drawn from our recent publications [69–73, 84, 103, 104]. Our results suggest the validity of a shift from the conventional wisdom of insurance practices that primarily serves the purpose of risk transfer to that of risk control and reduction; they also point to the validity of using insurance as an effective economic and incentive tool in cyber risk management.

1.4   VOLUNTARY PARTICIPATION

While clearly auto insurance is mandated for a vehicle owner, requiring insurance for other forms of risk is not always the case. To this end, the idea of making cyber insurance mandatory has mixed support due to concerns over geo-political factors and information asymmetry challenges in a networked world, and over the insurer's ability to precisely track and identify malicious actors and actions, thereby correctly assigning liability to concerned parties involved in a cyber insurance contract [31]. This is despite the increasing popularity of cyber insurance against the backdrop of increased cyber attacks on corporations around the globe since 2010, some of which have been mentioned earlier such as the Sony and Target data exfiltration attacks, and their accompanying negative impact on their business and reputation [32]. It remains to be seen whether cyber insurance will eventually become (semi-) mandatory, either through state mandate or through rigorous contractual relationship between firms and vendors.

With the above in mind, a general rule we will impose on ourselves in studies presented in this book is the notion of voluntary participation: we will assume that an agent is not required to purchase insurance. As we will see this assumption factors into our models and analysis very prominently. Voluntary cyber insurance brings with it the risk of bad matching of contracts between the supply and demand sides of insurance due to adverse selection challenges as noted earlier. In theory, we address this by assuming known risk types and known risk dependencies; in practice, this can only be dealt with empirically through methodical auditing processes.

1.5   RELATED LITERATURE

The growing cyber insurance market has motivated an extensive literature (see, e.g., [36, 59, 67, 68, 81, 90, 92, 94, 96, 101, 105–108, 125]), which aims to understand the unique characteristics of these emerging contracts, their effect on the insureds' security expenditure, and the possibility of leveraging these contracts to shape users' behavior and improve the state of cybersecurity.

The conclusions of these studies depend on the assumptions on the insurance market model (profit maker vs. welfare maximizing insurers), the agents' participation decisions (compulsory vs. voluntary insurance), and the assumed model of interdependency among the insured. Here we provide an overview of existing literature that is most closely related to this book.

Existing studies have considered either competitive or monopolistic insurers, as well as either mandatory or voluntary adoption by the insured. The works in [90, 105–108, 125] consider competitive insurance markets under compulsory insurance, and analyze the effect of insurance on agents' security expenditures. Shetty et al. [107, 108] consider a competitive market with homogeneous agents, and show that insurance often deteriorates the state of network security as compared to the no-insurance scenario. Schwartz et al. [105, 106] study a network of heterogeneous agents and show that the introduction of insurance cannot improve the state of network security.

Yang et al. [125] study a competitive market under the assumption of voluntary participation by agents, with and without moral hazard. In the absence of moral hazard, the insurer can observe agents' investments in security, and hence premium discriminates based on the observed investments. They show that such a market can provide incentives for agents to increase their investments in self-protection. However, they show that under moral hazard, the market will not provide an incentive for improving agents' investments.

The impact of insurance on the state of network security in the presence of a monopolistic welfare maximizing insurer has been studied in [37, 59, 67, 68, 95]. In these models, as the insurer's goal is to maximize social welfare, assuming compulsory insurance, agents are incentivized through premium discrimination. As a result, these studies show that insurance can lead to improvement of network security. An insurance market with a monopolistic profit maximizing insurer, under the assumption of voluntary participation, has been studied in [81], which shows that in the presence of moral hazard, insurance cannot improve network security as compared to the no-insurance scenario.

As mentioned earlier, one salient difference between a traditional insurance and cyber insurance lies in the fact that the propagating cyber risks are interdependent and statistically correlated, and is extremely challenging to get accurate estimates of. The challenge rapidly multiplies with an increase in the scale of the network. Methodologies to study the effect of interdependent cyber risks on the pricing of monopoly- and oligopoly-driven cyber insurance policies, and their impact on the overall security level of a network can be found in [81, 92, 94, 106, 107]. In particular, Ogut et al. [90] study the impact of the degree of agents' interdependence, and show that agents' investments decreases as the degree of interdependence increases. The main insight standing out from these works is that interdependent cyber risks, and information asymmetry between the cyber insurer and their clients fail to result in economically efficient cyber insurance markets that optimize cybersecurity. The effect of correlated cyber risks on the quality of cybersecurity induced by cyber insurance policies was studied in [35], and the main insight there is that a market for cyber insurance products is likely to exist under a network environment of correlated cyber risks; however, the quality of security it induces is sub-optimal, mainly because of inaccurate correlation estimates and information asymmetry.

Who could serve as the provider of this market mechanism has also been the subject of study, e.g., by a cyber insurer as those mentioned above, by a security vendor [93], through crowd funding [117], or through a coalitional insurance mechanism among organizations [118].

The next few chapters on cyber insurance are drawn from our recent publications [69–73]. We also refer the interested reader to [87] for an overview of contract theory.

1.6   ROADMAP

We will present in Chapter 2 a simple mathematical model describing the relationship between an insurer and an insured, and how the contract is determined between them. We will use this model to illustrate the importance of quantitative risk assessment, how it can effectively mitigate moral hazard, the difference it makes in determining the pricing of an insurance policy, and the impact it has on the subsequent cyber risk of the insured.

Then in Chapter 3 we take a deeper dive into this contract framework and explain why risk dependence, while certainly complicating matters, also emerges as an opportunity that can and should be exploited to the mutual benefit of both the insurer and the insured.

How the analysis and insights obtained in Chapter 3 translate into actual underwriting practices is described in Chapter 4, using a very common underwriting workflow process.

All of this is based on our ability to assess cyber risks at a firm level. We show in Chapter 5 a number of ways to quantify cyber risks in a global and scalable manner, and discuss the ensuing commercialization effort in this space and how our data analytics tools are being used today.

Chapter 6 outlines a number of open challenges in this domain and concludes the book.

Note on style and readability. We will use mathematical equations to describe models—we need the precision upon which to base our discussion. We will, however, largely avoid mathematical derivation or presenting formal, technical results, and opt for intuitive and qualitative explanations of these results instead. We provide all necessary references to the technical papers for an interested reader.

The point of view this book takes. There are many parties to an insurance policy: the enterprise/insured purchasing it, the underwriter selling it, the broker trying to match a buyer and a seller, the reinsurer providing insurance for the underwriter, and additional parties such as auditors and providers of risk and forensic analysis, among others. This book takes an outsider's holistic view of the most important part of this eco-system, i.e., the insured and the insurer and how their relationship impacts the state of cyber security for all. For this reason, we almost always consider the welfare of both parties, as well as a more objective measure of security. For simplicity of presentation, we will ignore the existence of the broker and roll into one the roles of insurer and reinsurer.

1.7   INTENDED AUDIENCE

This book is intended for someone interested in obtaining a quantitative understanding of cyber insurance and how innovation is possible around this centuries-old financial instrument. They broadly include the following.

1. Researchers working in the areas of cybersecurity, cyber risk quantification, and the economics of information security: they will be able to get a much better understanding of how cyber risk is quantified and managed in the context of insurance, and in particular, Chapters 4 and 5 will give them a very concrete sense of how it is practiced in reality. Such foundational knowledge should inspire ideas and open up new avenues of pursuit in their own research domains.

2. Researchers working in the area of mathematical economics and more specifically, incentive mechanisms: they may be interested in learning how insurance as a type of economic lever is used in the context of cybersecurity and its unique challenges, especially from Chapters 3 and 4.

3. Practitioners of cyber insurance on all sides, underwriters, brokers, reinsurers, and corporate risk managers who purchase insurance: some of them may not have a cybersecurity background, in which case they should find this book useful in providing some domain knowledge; some of them may be well-versed in both cybersecurity and (traditional) insurance products, in which case I hope the main premise of the book, that insurance could be used more effectively as a risk reducing mechanism, will give them fresh food for thought, and perhaps motivate them to adopt this line of thinking and seek similar innovation in their respective practices.

Some basic probability background would certainly help a reader get the most out of this book (risk after all is all about probability) but is by no means necessary. All technical results are accompanied by copious amount of explanation, so one could choose to skip the mathematical expositions and go straight for the plain language.

CHAPTER 2

A Basic Cyber Insurance Contract Model

In this chapter we will focus on the basic aspects of cyber insurance and present a few simple models to make our discussion precise and to illustrate the main insights. Our models serve to explain in mathematical terms the following concepts in the design of an insurance contract:

- a single-period contract vs. a multi-period contract;

- pre-screening vs. post-screening;

- what moral hazard and premium discrimination mean mathematically;

- what being risk averse vs. risk neutral means for the model; and

- the notions of individual rationality and incentive compatibility.

We will initially exclusively focus on the scenario of a single (risk-averse) agent (purchaser of the contract, or, to be used interchangeably, the insured) and a single (risk-neutral and profit-maximizing) principal (issuer of the contract, or, to be used interchangeably, the insurer or the underwriter). The analysis of the single-agent case allows us to study solely the role of *risk control* that a contract can play as part of the overall cyber risk management eco-system, by using tools such as pre-screening and post-screening. We will go on to investigate the situation of multiple, risk-correlated agents in the next two chapters to identify and explain the role of interdependency. A table summarizing notations used in this chapter can be found in Section 2.5.

2.1 A SINGLE-AGENT, SINGLE-PERIOD MODEL

Consider an agent, representing an organization (e.g., a business) with cyber risk exposures and who is potentially seeking insurance from a profit-maximizing insurer. We will use this simple scenario to construct a basic principal-agent model and highlight the concepts of moral hazard, risk aversion, and voluntary participation, in precise terms. To minimize confusion, we will generally use the pronoun she for the insurer and he for the insured.

2.1.1 THE MODEL

An illustration of the sequence of interactions between the insurer and the insured that our model will capture is given in Figure 2.1. We next specify each element of the model.

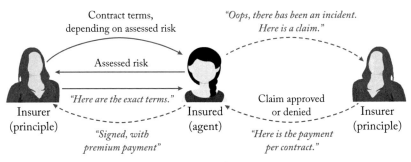

Figure 2.1: The interactions between the insurer and the insured.

Effort and cost of effort. The agent controls the amount of *effort* he can exert toward securing his system. This effort will be denoted as $e \in [0, +\infty)$, and it incurs a cost of c per unit of effort.

Loss. Let L_e denote the loss the agent may experience. This is in general a random variable, and the subscript reflects the idea that the amount of loss, while random, is influenced by his effort e—higher (more effective) effort tends to lead to lower losses. We will assume L_e has a mean of $\mu(e) \geq 0$ and variance $\lambda(e) \geq 0$.

While not always necessary, we will generally regard $\mu(e)$ and $\lambda(e)$ as strictly convex, strictly decreasing, and twice differentiable functions of e. The decreasing assumption implies that increased effort reduces the expected loss, as well as its unpredictability. The convexity assumption suggests that while initial investment in security leads to considerable reduction in loss, the marginal benefit decreases as effort increases. Indeed, it is generally not possible to reduce cyber risk to zero even if the agent exerts very large effort [62, 79].

We further assume that once the loss is realized (post-event), the exact amount is observed perfectly by both the insurer and the agent. This in practice is achieved through an auditing and forensics process.

Pre-screening. In general, the effort exerted by an agent is not observable by the insurer; this information asymmetry leads to the moral hazard phenomenon discussed earlier.

Suppose the insurer, in an effort to gain better information on the agent so as to come up with more informed policy terms, conducts a *pre-screening* of the agent's security standing, and as a result obtains an *assessment* or *outcome* in the form of $S_e = e + W$, where W is a noise term, assumed to have a zero mean with variance σ^2. We assume S_e is conditionally independent of L_e, given e. Later we will impose a more restrictive distributional assumption in order to be technically concrete, but that will not impact any of our qualitative conclusions.

In practice, the assessment S_e can be obtained through a number of methods and (Internet) measurement techniques, such as requiring agents to fill out surveys, external audits, or internal audits conducted by a third-party firm. We will devote Chapter 5 to this subject.

A linear contract and the insurer's payoff.   We consider the design of a set of *linear* contracts.[1] Specifically, it consists of a base premium $b \geq 0$, a discount factor α, and a coverage factor β. The agent pays a premium b, receives a discount $\alpha \cdot S_e$, and receives $\beta \cdot L_e$ as coverage in the event of a loss. We let $\alpha \geq 0$ so that the discount cannot become a penalty.[2] We will also let $0 \leq \beta \leq 1$, i.e., coverage never exceeds the actual loss. Thus, the insurer's utility (profit) is given by:

$$V(b, \alpha, \beta, e) = b - \alpha \cdot S_e - \beta \cdot L_e, \tag{2.1}$$

and her expected profit is given by

$$\overline{V}(b, \alpha, \beta, e) = E[V(b, \alpha, \beta, e)] = b - \alpha e - \beta \mu(e). \tag{2.2}$$

A risk-averse agent.   The utility of a risk-averse agent, if he does not purchase an insurance contract, is given by (the negative of) his total loss:

$$U(e) = -\exp\{-\gamma \cdot (-L_e - ce)\}, \tag{2.3}$$

where γ denotes the *risk attitude* of the agent; a higher γ implies more risk aversion. We assume γ is known to the insurer, thereby eliminating adverse selection and solely focusing on the moral hazard aspect of the problem. In other words, if the agent chooses not to enter a contract, he bears the full cost of his effort as well as any realized loss.

The expected utility of the agent outside a contract is then $\overline{U}(e) = E[-\exp\{-\gamma \cdot (-L_e - ce)\}]$. Therefore, the optimal effort of the agent outside the contract is

$$e^o = \arg\max_{e \geq 0} \overline{U}(e), \tag{2.4}$$

and his expected utility outside the contract is denoted by $u^o := \overline{U}(e^o)$.

If this agent accepts a contract (b, α, β), his utility is given by:

$$U^{in}(b, \alpha, \beta, e) = -\exp\{-\gamma \cdot (-b + \alpha \cdot S_e - L_e + \beta \cdot L_e - ce)\}.$$

A risk-neutral agent.   For completeness and for later comparison, we will also introduce a risk-neutral agent. In order not to overload our notation we will reuse the same symbols while making it clear which type of agent is under consideration. The risk-neutral agent's utility is also given by (the negative of) his total loss if he chooses not to enter a contract:

$$U(e) = -L_e - ce \Rightarrow \overline{U}(e) = E[U(e)] = -\mu(e) - ce. \tag{2.5}$$

---

[1]As with other technical assumptions, this particular contract form is considered for simplicity of exposition. There are many ways to refine this model, but they do not impact the qualitative conclusions of our analysis.

[2]This is without loss of generality as we can show that under an optimal contract this factor cannot be negative; see the proof of Lemma 2.4 in Section 2.6.

His optimal effort (e^o) outside the contract is

$$e^o = \arg\min_{e \geq 0} \; \mu(e) + ce, \tag{2.6}$$

and his expected utility outside the contract is $u^o := \overline{U}(e^o)$. If the agent purchases a contract (p, α, β), then his utility is given by:

$$U^{in}(b, \alpha, \beta, e) = -b + \alpha S_e - L_e + \beta L_e - ce. \tag{2.7}$$

The insurer's problem.   The insurer designs the contract (b, α, β) to maximize her expected payoff. In doing so, the insurer also has to satisfy two standard constraints: Individual Rationality (IR) and Incentive Compatibility (IC). The first stipulates that a rational agent will not enter a contract with expected payoff less than his outside option u^o, since participation in insurance is voluntary. The second stipulates that the insurer in designing the contract assumes the agent will act rationally (out of self-interest), in the sense that the effort exerted by the agent under the contract should maximize the agent's expected utility under that contract. Formally,

$$\max_{b, e, \alpha \geq 0, 0 \leq \beta \leq 1} \quad \overline{V}(b, \alpha, \beta, e) \tag{2.8}$$

$$\text{s.t.} \quad \text{(IR)} \quad \overline{U}^{in}(b, \alpha, \beta, e) \geq u^o$$

$$\text{(IC)} \quad e \in \arg\max_{e' \geq 0} \; \overline{U}^{in}(b, \alpha, \beta, e').$$

In the above maximization, while it appears that the insurer is also selecting the best e, she does so because she expects the agent to do so by following the IC constraint; thus, she merely imitates what she believes the agent will do.

Who knows what and when.   Since this is a principal-agent problem with a first mover and a second mover, it is instructive to spell out the sequencing of actions and the informational structure. First of all, values such as the cost of effort c, the agent's risk attitude γ, the distribution of loss L_e as a function of e, and the insurer's pre-screening accuracy (the distribution of W, or the noise σ^2) are all public knowledge. Second, the format of the policy, i.e., that it consists of a premium b, a discount on premium α, and a coverage factor β, while dictated by the insurer, is also public knowledge.

The sequence of actions and events that follow is illustrated in Figure 2.2. As highlighted:

1. Both the insurer and the agent solve the constrained optimization problem (2.8), obtaining the optimal values b, α, β, and e. This simultaneously solves the insurer's and the agent's problems, the optimal contract terms (b, α, β) for the former, and the optimal effort level e for the latter.

   Note that problem (2.8) is guaranteed to have a solution since a contract with terms $(0, 0, 1)$ would obviously satisfy the IR constraint; thus, the result is always acceptable to the agent.

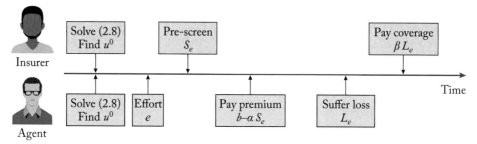

Figure 2.2: The sequence of actions in the design and execution of the contract.

On the other hand, the insurer may not find the resulting contract profitable, i.e., it is possible that $\overline{V} < 0$ even for the best contract, in which case the insurer will not offer the contract and walk away from the agent. Let's assume the contract is offered (and thus accepted) and proceed.

2. The agent exerts effort e because this is the incentive compatible action.

3. The insurer conducts pre-screening and obtains S_e.

4. The agent pays a total premium less discount of $b - \alpha S_e$; the contract is now active.

5. If a loss event occurs during the contract term in the amount L_e, the agent receives coverage βL_e.

It may seem that the problem (2.8) is solved by both parties simultaneously, so why do we call the insurer the first mover and the agent the second? This is because what's implied is the fact that the former has decided on the format of the contract: the triple (b, α, β), and she announces this to the latter, who then best responds (with e). But since e falls out of the solution to (2.8), the insurer anticipates it perfectly, just as the agent has the ability to anticipate perfectly the values (b, α, β). So in essence, both sides effectively solve (2.8) simultaneously.

We will now use this model to highlight the following concepts: (1) the difference between a risk-neutral agent and a risk-averse agent; (2) what risk aversion means to the optimal contract and optimal effort, and why we call the insurance a risk transfer instrument; and (3) what impact does the insurer's pre-screening ability have on the agent's effort, her own utility, and in our application context, the state of security.

2.1.2   RISK NEUTRALITY VS. RISK AVERSION

To proceed, we will now assume that both L_e and W are of a normal distribution: L_e of mean $\mu(e) \geq 0$, variance $\lambda(e) \geq 0$,[3] and W of 0 mean and variance σ^2 as noted earlier. The normal

---

[3]For ease of exposition, we assume that $\lambda(e)$ is sufficiently small compared to $\mu(e)$, so that $Pr(L_e < 0)$ is negligible.

assumption on L_e is to some extent justified by the fact that L_e is meant to capture the sum total of losses from a variety of sources, such as hacking, malware, insider threats, etc. But mostly it is adopted here to simplify our derivation so we do not get distracted from the main observations, which do not rely on this technical assumption. We are now ready to solve (2.8) for both types of agent.

The IR constraint is binding (held with equality) under the optimal contract. We will not provide a formal proof for this first result, but the intuition should be quite obvious: if the equality does not hold and $\overline{U}^{in} > u^o$, then we can always increase the premium b by a tiny amount without violating the IR constraint. At the same time, this increase also does not impact the IC constraint as the choice of the optimal effort e is independent of b (this is seen more concretely shortly, e.g., in (2.10)). This suggests we have found a new feasible solution. Therefore by suitably increasing b the insurer can improve her profit $\overline{V}$, which is a contradiction to the original contract being optimal. Therefore, if the contract is indeed optimal, the two sides of the IR constraint must be equal.

For this reason, the optimal premium b can be recovered from the IR constraint, and we will frequently suppress this argument in writing the insurer's profit as $V(\alpha, \beta)$.

Risk-neutral agent. The binding IR constraint means that

$$-b - (c - \alpha) \cdot e - (1 - \beta)\mu(e) = u^o. \tag{2.9}$$

Using (2.9) to substitute for the base premium b in (2.2), we see that the insurer seeks to maximize (over e) her expected utility $\overline{V} = -u^o - \mu(e) - ce$. At the same time, the agent's outside utility u^o is given by $u^o = \max_{e \geq 0}\{-\mu(e) - ce\}$. This immediately leads to the observation that the insurer's profit is *at most zero*. In particular, a contract with ($b = 0, \alpha = 0, \beta = 0$) will yield a payoff of zero, making it an optimal contract: no premium, no discount, and no coverage!

We thus conclude that the insurer would not offer a contract to a risk-neutral agent at all. Furthermore, in this case the quality of pre-screening, or indeed the availability of pre-screening regardless of the quality, plays no role in either the insurer's or agent's decisions.

This is why we say risk aversion is what drives people to purchase insurance: with a risk-neutral agent, there is no profitable contract to sell and the market would not exist.

Risk-averse agent. For a risk-averse agent, his expected utility is given as follows, noting the conditional independence between S_e and L_e:

$$
\begin{aligned}
\overline{U}^{in}(b, \alpha, \beta, e) &= E[-\exp\{-\gamma \cdot (-b + \alpha \cdot S_e - L_e + \beta \cdot L_e - ce)\}] \\
&= -\exp\{\gamma(b + (c - \alpha)e + \frac{\gamma}{2}\alpha^2\sigma^2 + (1 - \beta)\mu(e) + \frac{\gamma}{2}(1 - \beta)^2\lambda(e))\}.
\end{aligned}
$$

Therefore, the corresponding IC constraint for this risk-averse agent is:

$$e \in \arg\min_{e' \geq 0}\{(c - \alpha)e' + (1 - \beta)\mu(e') + \frac{\gamma}{2}(1 - \beta)^2\lambda(e')\}. \tag{2.10}$$

The binding IR constraint means

$$b + (c - \alpha)e + \frac{\gamma}{2}\alpha^2\sigma^2 + (1 - \beta)\mu(e) + \frac{\gamma}{2}(1 - \beta)^2\lambda(e) = m^o, \tag{2.11}$$

where $m^o := \frac{\ln(-u^o)}{\gamma}$ for convenience. Note that outside the contract, we have $e^o = \min_{e \geq 0}\{\mu(e) + \frac{\gamma}{2}\lambda(e) + c \cdot e\}$. We can thus rewrite the contract design problem (2.8) as follows:

$$\max_{e,\alpha \geq 0, 0 \leq \beta \leq 1} \quad m^o - \mu(e) - \frac{\gamma}{2}(1 - \beta)^2\lambda(e) - ce - \frac{\gamma}{2}\alpha^2\sigma^2$$

$$\text{s.t.} \quad (\text{IC}) \quad e = \arg\min_{e' \geq 0}\{(c - \alpha)e' + (1 - \beta)\mu(e') + \frac{\gamma}{2}(1 - \beta)^2\lambda(e')\}. \tag{2.12}$$

We can now solve the optimal contract problem posed in (2.12). The first result below shows that not only is there now a profitable contract to sell, but also that the agent's effort inside the contract is at or lower than that outside the contract.

Theorem 2.1   *Assume that (α, β, e) solves optimization problem (2.12). Then $e \leq e^o$, where e^o is the optimal level of effort outside the contract; in other words, insurance decreases network security.*

Proof. Assume that (α, β, e) solves optimization problem (2.12), and that, by contradiction, $e > e^o \geq 0$.

First, recall $e^o := \arg\min_{e \geq 0}\{\mu(e) + \frac{1}{2}\gamma\lambda(e) + ce\}$. For e^o to be the minimizer, we should have $c + \mu'(e^o) + \frac{1}{2}\gamma\lambda'(e^o) \geq 0$. Next, consider the following two cases.

(1) $\alpha = 0$. Starting from the first order condition (FOC) on the IC constraint, we have:

$$c + (1 - \beta)\mu'(e) + \frac{1}{2}\gamma(1 - \beta)^2\lambda'(e) = 0$$

$$\Rightarrow \quad c + \mu'(e) + \frac{1}{2}\gamma\lambda'(e) < 0$$

$$\Rightarrow \quad c + \mu'(e^o) + \frac{1}{2}\gamma\lambda'(e^o) < 0, \tag{2.13}$$

where the first inequality follows from the decreasing nature of $\mu(\cdot)$ and $\lambda(\cdot)$, and the second inequality follows from their convexity. However, the second inequality is impossible given the optimality of e^o outside the contract (the FOC dictates that this should be an equality). This contradiction shows that we cannot have $e > e^o$.

(2) $\alpha > 0$. Given the assumption that $e > e^o$, and $\mu(\cdot)$ and $\lambda(\cdot)$ are strictly convex and decreasing, we have:

$$
\begin{aligned}
0 &= c + \mu'(e^o) + \frac{1}{2}\gamma\lambda'(e^o) \\
&\leq c + \mu'(e^o) + \frac{1}{2}\gamma(1-\beta)^2\lambda'(e^o) \\
&< c + \mu'(e) + \frac{1}{2}\gamma(1-\beta)^2\lambda'(e).
\end{aligned}
\tag{2.14}
$$

This immediately suggests that the insurer's utility function value under the optimal contract, given by $m^o - \mu(e) - \frac{1}{2}(1-\beta)^2\lambda(e) - ce - \frac{1}{2}\gamma\alpha^2\sigma^2$ will increase if we decrease e or decrease α. At the same time, we know from the IC constraint that if the insurer deviates from this contract and decreases α, the agent will best respond by decreasing e. Therefore, by decreasing α the insurer can increase her utility, which contradicts the optimality of (α, β, e). Thus, by contradiction we conclude that the agent's effort in the optimal contract cannot exceed e^o. $\quad\square$

Theorem 2.1 illustrates two things. First, when an agent is risk averse, the insurance market can exist, meaning there is a profitable contract to sell and the agent will accept it. Second, it shows the inherent inefficiency in using cyber insurance as a tool to improve the state of security as it causes the agent to lower his effort (note also that Theorem 2.1 holds regardless of the pre-screening quality). The same observations have been made under various models, e.g., see works in [95, 107] when studying competitive and/or unregulated cyber insurance markets.

2.1.3 THE ROLE OF PRE-SCREENING

We next examine the role of pre-screening under this model. We first analyze its impact on the insurer's profit.

Theorem 2.2 *The insurer's optimal payoff is a decreasing function of her pre-screening noise σ^2.*

Proof. Let $V(\alpha, \beta, e, \sigma^2)$ denote the principal's payoff with contract (α, β) when the agent exerts effort e, and pre-screening noise σ^2. Let $z(\sigma^2)$ denote the maximum payoff as a function of σ^2. Below we show that $z(\sigma_1^2 + \sigma_2^2) \leq z(\sigma_1^2)$, $\forall \sigma_2^2$.

Again the insurer's utility function is given by (from (2.12)):

$$
V(\alpha, \beta, e, \sigma^2) = m^o - \mu(e) - \frac{\gamma}{2}(1-\beta)^2\lambda(e) - ce - \frac{\gamma}{2}\alpha^2\sigma^2,
\tag{2.15}
$$

where m^o does not depend on any of the contract parameters. Thus, we have

$$
\begin{aligned}
z(\sigma_1^2 + \sigma_2^2) &= \max_{0 \leq \beta \leq 1, e, \alpha \geq 0, \text{IC}} V(\alpha, \beta, e, \sigma_1^2 + \sigma_2^2) \\
&\leq \max_{0 \leq \beta \leq 1, e, \alpha \geq 0, \text{IC}} V(\alpha, \beta, e, \sigma_1^2) + \max_\alpha \{-\frac{\gamma}{2}\alpha^2 \sigma_2^2\} \\
&\leq \max_{0 \leq \beta \leq 1, e, \alpha \geq 0, \text{IC}} V(\alpha, \beta, e, \sigma_1^2) = z(\sigma_1^2). \quad (2.16)
\end{aligned}
$$

Therefore, the insurer's optimal payoff $z(\sigma^2)$ is a decreasing function of the pre-screening noise.

$\square$

The above result says the following: a strategic insurer can leverage improved pre-screening to better mitigate moral hazard and attain a higher payoff, which is to be expected.

Second, (2.15) suggests that as the pre-screening becomes less accurate (higher σ^2) the optimal contract will place less emphasis on premium discount (smaller α). In the extreme case of $\sigma^2 = \infty$, we have $\alpha = 0$ in an optimal contract, effectively eliminating discount and pre-screening. For this reason, we will let $\sigma = \infty$ denote the scenario of no pre-screening.

The next result examines the effect of the quality of pre-screening on the state of security, and shows that under a suitable sufficient condition, the availability of pre-screening assessment improves the agent's effort, compared to no pre-screening, and the more accurate the pre-screening, the higher the agent's effort.

Theorem 2.3 *Let e_1, e_2, e_∞ denote the optimal effort of the agent in the optimal contract when $\sigma = \sigma_1, \sigma = \sigma_2$ and $\sigma = \infty$, respectively. Let $k(e, \alpha) = \frac{\mu'(e) + \sqrt{\mu'(e)^2 - 2\gamma(c-\alpha)\lambda'(e)}}{-\gamma\lambda'(e)}$. If $k(e, \alpha_1)^2\lambda(e) - k(e, \alpha_2)^2\lambda(e)$ is non-decreasing in e for all $0 \leq \alpha_1 \leq \alpha_2 \leq c$, then $e_1 \geq e_2$ if $\sigma_1 \leq \sigma_2$. In addition, if $k(e, 0)^2\lambda(e) - k(e, \alpha)^2\lambda(e)$ is non-decreasing in e for all $0 \leq \alpha \leq c$, then $e_1 \geq e_\infty$.*

Several instances of $\mu(e)$ and $\lambda(e)$, e.g., $(\mu(e) = \frac{1}{e}, \lambda(e) = \frac{1}{e^2})$, and $(\mu(e) = \exp\{-e\}, \lambda(e) = \exp\{-2e\})$, satisfy the condition of Theorem 2.3. The proof can be found in Appendix 2.6.

2.1.4 SUMMARY

We summarize the main takeaways from this section. By comparing the contracts in the risk-neutral and risk-averse agent cases, we observe that a market exists and the insurer makes profit only when offering a contract to a risk-averse agent. This is indeed to be expected, as insurance is primarily a method for risk transfer; risk-averse agents are willing to pay premiums that are higher than their expected loss, in order to reduce the uncertainty in their loss, consequently allowing the risk-neutral insurer to make a profit. We further observe that when the market exists, the introduction of pre-screening benefits the insurer (Theorem 2.2) as well the state of network security (Theorem 2.3).

2.2   A SINGLE-AGENT, MULTI-PERIOD MODEL

The previous section focuses on a single contract period, where the terms of the contract and the conditions and actions of the insured are independent of any history and without anticipation of the future. In practice, we could view pre-screening as assessing present conditions that summarize past history (i.e., that pre-screening could generate a *sufficient statistic*), so history is not really ignored. On the other hand, the ability to anticipate future can significantly alter decisions and actions. This motivates the consideration of a multi-period model, where the terms of a policy could depend on what happened in the past, and a policy holder's strategic decisions will take into account future consequences of their current actions.

Post-screening.   Taking history into consideration in determining the contract terms for the next period may be viewed as a form of *post-screening*, to contrast with *pre-screening*. Post-screening is very commonly done in practice; e.g., a speeding ticket or automobile accident could lead to increased premium when the policy is renewed. In this section we will present a multi-period contract model. For our purposes it is sufficient to consider a two-period model, which can be generalized to an n-period model.

2.2.1   A TWO-PERIOD CONTRACT MODEL

So as not to overload our notation, in this two-period setting we will assume losses are fully covered (paid in full), thereby dropping the β component (or equivalently, letting $\beta = 1$). While we could carry on without this omission, the simplification allows us to get to the essence of our analysis more straightforwardly without affecting the main qualitative conclusions.

Second, we will assume that the loss L_e is a binary random variable: a constant loss amount of L is incurred with probability p_e, assumed to be strictly decreasing and strictly convex in e, and with probability $1 - p_e$ nothing happens, implying that the initial effort toward security leads to a considerable reduction in the probability of a loss incident but the marginal benefit decreases.

The expected utility of the agent outside insurance is given by:

$$
\begin{aligned}
\overline{U}(e) &= E[-\exp\{-\gamma(-L_e - ce)\}] \\
&= p_e(-\exp\{-\gamma(-L - ce)\}) + (1 - p_e)(-\exp\{-\gamma(-ce)\}),
\end{aligned} \tag{2.17}
$$

with risk attitude γ; again, the higher the risk attitude the more risk averse the agent. We will again denote by $u^o := \max_e \overline{U}(e)$ the agent's maximum utility outside the contract.

2.2.2   TWO-PERIOD CONTRACT DESIGN WITH POST-SCREENING

Since the insurer is able to assess premium in the second period based on what happened in the first period, such a contract is given by the triple (b_1, b_2, b_3): b_1 is the first-period premium; in the second period, the agent pays premium b_2 if a loss happened during the first period and pays b_3 otherwise. Obviously, $b_3 \leq b_2$.

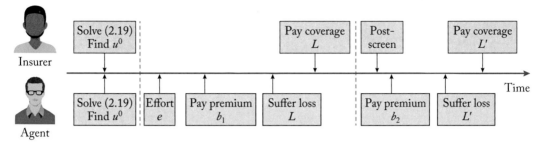

Figure 2.3: The sequence of actions in the design and execution of the two-period contract; dashed lines delineate the two successive policy periods.

In this case, the agent may exert non-zero effort in the first period to decrease the chance of a loss in order to reduce the likelihood of paying a higher premium in the second period. In the second period, on the other hand, the agent will always exert zero effort as the loss is fully covered and he faces no more future punishment.[4] For this reason, the probably of loss during the second period is given by p_0.

We assume that when an agent enters such a contract he commits to both periods. The agent's utility inside a contract (b_1, b_2, b_3) with post-screening is thus the summation of his utility in each period:

$$\overline{U}^{in}(e, b_1, b_2, b_3) = g(-b_1 - ce) + p_e g(-b_2) + (1 - p_e)g(-b_3), \qquad (2.18)$$

where e is the effort in the first period, and recall $g(x) = -\exp\{-\gamma x\}$.

As before, the insurer's problem is to maximize her profit subject to the IR constraint and IC constraint:

$$\overline{V} = \max_{e, b_1, b_2, b_3 \geq 0} \quad b_1 - p_e L + p_e(b_2 - p_0 L) + (1 - p_e)(b_3 - p_0 L) \qquad (2.19)$$

$$\text{s.t.} \quad \text{(IR)} \quad \overline{U}^{in}(e, b_1, b_2, b_3) \geq 2 \cdot u^o, \ i = 1, 2,$$

$$\text{(IC)} \quad e \in \arg\max_{e'} \overline{U}^{in}(e', b_1, b_2, b_3).$$

The sequence of actions and events over these two periods is illustrated in Figure 2.3.

Recall that the IR constraint ensures that the agent enters the contract only if he gets no lower utility than his outside option. Since the contract covers two periods, the comparison here is between his utility inside the contract over two periods and outside the contracts over two periods.

---

[4]This extends naturally to a multi-period setting where the premium of each period depends on the agent's history of losses. For instance, the agent's third-period premium depends on his loss events in the first and second periods, and so on.

Under the contract (b_1, b_2, b_3), by the first-order optimality condition, the agent's optimal effort is given by:

$$e(b_1, b_2, b_3) = \begin{cases} \left(\frac{1}{\alpha + \gamma c} \ln(t \cdot \frac{\alpha}{\gamma c} \frac{\exp\{\gamma b_2\} - \exp\{\gamma b_3\}}{\exp\{\gamma b_1\}}) \right)^+ & \text{if } b_2 > b_3 \\ 0 & \text{if } b_2 \leq b_3 \end{cases}. \tag{2.20}$$

It can be shown using similar techniques as before that the IR constraint in the optimization problem (2.19) is binding; thus, at the optimal solution the agent is indifferent between entering vs. not entering the contract, as expected.

2.2.3   TWO-PERIOD CONTRACT DESIGN WITH PRE-SCREENING

We now turn to the case of pre-screening, where the insurer can conduct a risk assessment prior to determining the contract terms. As before, the outcome of the pre-screening is given by an assessment $S_e = e + W$, where W is a zero-mean Gaussian noise with variance σ^2.

The insurer then offers the agent a contract given by two parameters (b, α), where b is the base premium and α the assessment-dependent discount factor: the agent pays $b - \alpha S_e$ in exchange for full coverage in the event of a loss. This problem is now identical to the one presented in Section 2.1 by setting $\beta = 1$. The agent's total cost inside the contract (b, α) while exerting effort e is $b - \alpha S_e + ce$, which follows a Gaussian distribution. Thus, using moment-generating function the agent's expected utility under the contract is given by:

$$\overline{U}(b, \alpha, e) = E[g(-(b - \alpha S_e + ce))] = -\exp\{\gamma b + \gamma(c - \alpha)e + \tfrac{\gamma^2 \alpha^2 \sigma^2}{2}\}. \tag{2.21}$$

Therefore, the insurer's design problem using pre-screening is as follows:

$$\max_{\alpha, b, e \geq 0} \quad E[b - \alpha S_e] - p_e L$$

$$\text{s.t.} \quad \text{(IR)} \quad \overline{U}(b, \alpha, e) \geq u^o$$

$$\text{(IC)} \quad e \in \arg\max_{e' \geq 0} \overline{U}(b, \alpha, e'). \tag{2.22}$$

Similar as before, we can show that the IR constraint is binding in this case. Thus, we have the following relationship between the optimal contract parameters:

$$b = m^o + \alpha e - ce - \frac{\gamma \alpha^2 \sigma^2}{2}, \tag{2.23}$$

where again $m^o = \frac{1}{\gamma} \ln(u^o)$. Using (2.23), the insurer's problem can be simplified as follows:

$$\max_{\alpha, e \geq 0} \quad m^o - ce - \frac{\gamma \alpha^2 \sigma^2}{2} - p_e L$$

$$\text{s.t.} \quad \text{(IC)} \quad e \in \arg\min_{e' \geq 0}\{(c - \alpha)e' + \frac{\gamma \alpha^2 \sigma^2}{2}\}. \tag{2.24}$$

2.2.4 STATE OF SECURITY AND OPTIMAL CONTRACTS

We briefly summarize known results on these two types of premium discrimination in terms of their effectiveness in incentivizing efforts.

Post-screening: Post-screening has been studied in the literature. Rubinstein et al. [102] showed that post-screening can improve the agent's effort inside the contract compared to the one-period contract without post-screening. This can be similarly observed in our model, where a sufficient condition can be established under which the agent exerts non-zero effort in the first period of a contract with post-screening [74].

Pre-screening: The result in Section 2.1 shows that pre-screening can simultaneously incentivize the agent to exert non-zero effort and improve the insurer's utility. It turns out, that when losses are rare and are perceived differently by the insurer and by the agent, pre-screening can incentivize higher effort whereas post-screening cannot [74].

For the remainder of this book, we will primarily use pre-screening and thus limit our model to a single period. We will, however, discuss repeated use of pre-screening in a multi-period model in Chapter 6.

2.3 NUMERICAL RESULTS

We present some numerical examples of the findings presented in this chapter. Our main focus is on the impact of pre-screening noise in various scenarios. Throughout the first part of this section we will use the following parameters:

$$\mu(e) = \frac{10}{e+1}, \quad \lambda(e) = \frac{10}{(e+1)^2}, \quad c = 2, \quad \gamma = 1. \tag{2.25}$$

2.3.1 IMPACT OF AGENT'S RISK ATTITUDE γ

Figure 2.4 illustrates the optimal contract as a function of γ. As the agent becomes more risk averse, the insurer can set higher premium b and lower discount α, while offering higher coverage β.

Figure 2.5 illustrates network security (agent's effort), both inside and outside a contract, vs. his risk attitude γ. First, we see that as suggested by Theorem 2.1, the agent's effort under the contract is less than that outside the contract: insurance decreases network security as the agent transfers his risk to the insurer thereby removing his own incentive to exert high effort. We also observe that the agent's effort under the optimal contract is a decreasing function of γ. This is because as the agent becomes more risk-averse, he transfers more risk to the insurer (seen in Figure 2.4) and further decreases his own effort. Indeed, for this reason pre-screening also becomes less important as the agent's risk-aversion increases, making their own efforts less important. Finally, when the agent is outside the contract, he can only decrease his risk by ex-

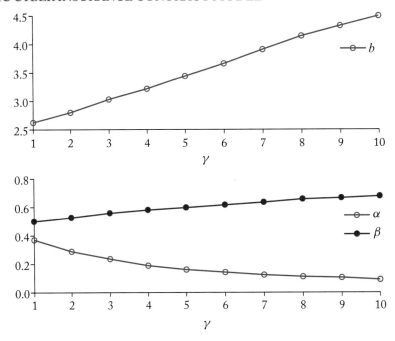

Figure 2.4: Parameters of the optimal contract vs. risk aversion level γ.

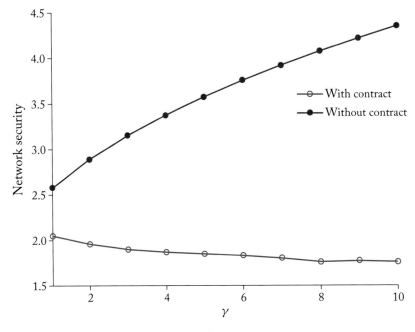

Figure 2.5: Effort of agent vs. risk aversion level γ.

erting higher effort. Therefore, we observe that as an agent without insurance becomes more risk-averse, he exerts higher effort.

2.3.2   IMPACT OF PRE-SCREENING NOISE

Figure 2.6 illustrates the insurer's profit as a function of the pre-screening noise σ^2. The observation is consistent with Theorem 2.2, which states that the insurer's profit is a decreasing function of σ^2. Figure 2.7 illustrates the effort of the agent inside and outside the contract as a function of σ^2. We see that the effort outside the contract is independent of the pre-screening noise, while it decreases inside the contract as σ^2 increases. This highlights the fact that as the insurer becomes less accurate in her observation of the agent's effort, she starts to place less importance on the pre-screening outcome. As a result, it becomes less beneficial for the agent to exert high effort without receiving sufficient discount. In other words, low-quality pre-screening dampens its effectiveness in mitigating moral hazard; consequently, network security worsens. A second observation here is that as the participation constraint is always binding, the constant effort outside the contract also means that the agent's utility remains constant regardless of the pre-screening noise. Thus, it is only the insurer who benefits from pre-screening.

2.3.3   EXPONENTIAL LOSS AND UNIFORM PRE-SCREENING NOISE

Throughout our analysis, we assumed that losses and pre-screening outcomes are normally distributed. In this section, we provide a numerical example under the assumption of exponentially distributed losses and uniformly distributed pre-screening outcomes. We illustrate how our previous observations hold in this instance as well. Consider:

$$
\begin{aligned}
&\gamma = 0.9,\ c = 0.25,\ E(L_e) = \mu(e) = \tfrac{1}{1+e}, \\
&L_e \sim \exp(\tfrac{1}{\mu(e)}), \\
&S_e = e + W,\ W \sim \mathrm{Unif}(-w, w).
\end{aligned}
\tag{2.26}
$$

Figure 2.8 illustrates the agent's effort in this case. We see that the change in loss and pre-screening noise assumptions did not change our key observations, that the agent's effort inside the contract is less than outside the contract and that it remains a decreasing function of the magnitude of the noise.

2.4   CHAPTER SUMMARY

This chapter introduces a basic insurance model involving a single insurer and a single risk-averse agent, over single or multiple contract periods. Analysis of this model highlights the extent to which a contract can control the risk of the insured through effective premium discrimination enabled by accurate pre-screening. In the next chapter we will build on this model to investigate the case of multiple, risk-correlated agents.

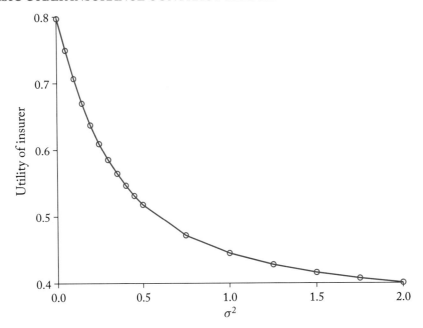

Figure 2.6: Insurer's profit vs. pre-screening noise σ^2.

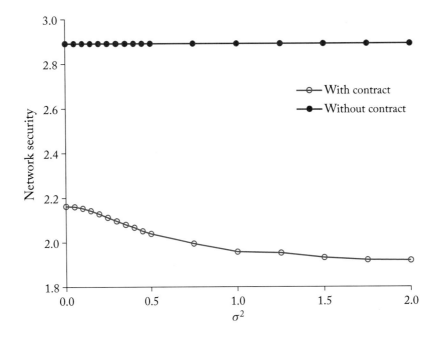

Figure 2.7: Agent's effort vs. pre-screening noise σ^2.

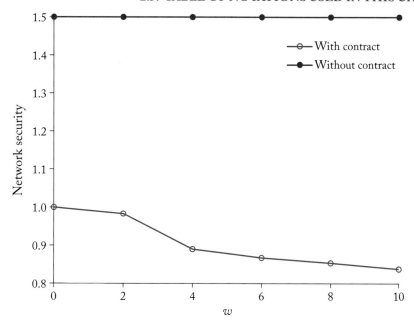

Figure 2.8: Agent's effort vs. pre-screening noise with exponentially distributed loss.

2.5   TABLE OF NOTATIONS USED IN THIS CHAPTER

See Table 2.1.

2.6   APPENDIX

2.6.1   PROOF OF THEOREM 2.3

The following lemma is used in the proof of Theorem 2.3.

Lemma 2.4   *In an optimal contract, $0 \leq \alpha \leq c$.*

Proof. Assume (α, β, e) is the optimal solution to (2.12). The Karush–Kuhn–Tucker (KKT) condition for the IC constraint is as follows:

$$(1 - \beta)\mu'(e) + \frac{1}{2}\gamma(1 - \beta)^2\lambda'(e) - v \; = \; \alpha - c, \tag{2.27}$$
$$v \cdot e \; = \; 0, \quad v, e \geq 0.$$

In the above equation, the left-hand side is negative, as both $\mu(\cdot)$ and $\lambda(\cdot)$ are decreasing, and the slack variable $v \geq 0$. Therefore, we must have $\alpha \leq c$ from the first equation.

Table 2.1: Table of notations used in this chapter

Symbol	Definition
e, c	Agent's effort in security and its unit cost
γ	Agent's risk attitude
$L_e, \mu(e), \lambda(e)$	Agent's loss, its mean, and variance
S_e, W, σ^2	Agent's effort assessed by the insurer, noise in the estimate, and its variance
b	Premium of a policy
α	Discount factor
β	Coverage factor
$V(b, \alpha, \beta, e), \overline{V}$	Insurer's utility and its expectation
$U(e), \overline{U}$	Agent's utility outside insurance and its expectation
e^o	Agent's optimal effort outside insurance
u^o	Agent's optimal expected utility outside insurance
$U^{in}(b, \alpha, \beta, e), \overline{U}^{in}$	Agent's utility under insurance and its expectation
$m^o := \dfrac{\ln(-u^o)}{\gamma}$	Introduced for notational convenience
p_e, L	Probability and amount of loss in the binary loss model
(b_1, b_2, b_3)	Premium values of a two-period contract
$g(x) = -\exp\{-\gamma x\}$	Introduced for notational convenience
$e(b_1, b_2, b_3)$	Agent's optimal effort under two-period contract (b_1, b_2, b_3)

Since we have assumed $\alpha \geq 0$ all along, the proof of this lemma is technically complete. However, below we show that under an optimal contract α cannot be negative even if we had allowed it. To prove this, let us assume that $\alpha < 0$. In this case, we must have $v = 0$; otherwise if $v > 0$, then we can decrease v and increase α by the same amount (while still keeping $\alpha < 0$) in the KKT condition (2.27) without changing β and e. This would increase the objective function of the insurer (since we have decreased α^2), contradicting the optimality of the solution. Therefore, if $\alpha < 0$, we must have $v = 0$.

Setting $v = 0$ in (2.27), we have:

$$c + (1 - \beta)\mu'(e) + \frac{1}{2}\gamma(1 - \beta)^2\lambda'(e) = \alpha < 0. \tag{2.28}$$

From this, we get

$$-\mu'(e) - \frac{1}{2}\gamma(1-\beta)^2\lambda'(e) - c \geq -c - (1-\beta)\mu'(e) - \frac{1}{2}\gamma(1-\beta)^2\lambda'(e) > 0. \qquad (2.29)$$

That is, $-\mu(e) - \frac{1}{2}\gamma(1-\beta)^2\lambda(e) - ce$ is an increasing function of e.

Next, note that as $\mu(\cdot)$ and $\lambda(\cdot)$ are convex, we can find $0 > \alpha' > \alpha$ and $e' > e$ in (2.28) such that

$$c + (1-\beta)\mu'(e') + \frac{1}{2}\gamma(1-\beta)^2\lambda'(e') = \alpha'. \qquad (2.30)$$

Taking together the facts that $-\mu(e) - \frac{1}{2}\gamma(1-\beta)^2\lambda(e) - ce$ is increasing in e, $e' > e$, and $0 > \alpha' > \alpha$, the contract (α', β, e') improves the objective function value of the insurer compared to (α, β, e), contradicting the optimality of the latter. From this contradiction, we conclude that α cannot be negative. $\qquad \square$

We are now ready to prove Theorem 2.3.

Assume $\sigma_1^2 \leq \sigma_2^2$. Let (α_i, β_i) and e_i be the parameters of optimal contract and the optimal effort of the agent in that contract, respectively, when the pre-screening noise is $\sigma^2 = \sigma_i^2$.

First we show that $\alpha_2 \leq \alpha_1$ by contradiction. Assume instead $\alpha_2 > \alpha_1$. From the optimality of (α_1, β_1) when $\sigma = \sigma_1$, we have

$$-\mu(e_1) - \frac{1}{2}\gamma(1-\beta_1)^2\lambda(e_1) - ce_1 - \frac{1}{2}\gamma\sigma_1^2\alpha_1^2$$
$$\geq -\mu(e_2) - \frac{1}{2}\gamma(1-\beta_2)^2\lambda(e_2) - ce_2 - \frac{1}{2}\gamma\sigma_1^2\alpha_2^2. \qquad (2.31)$$

In addition, since $\alpha_2 > \alpha_1$, we have

$$\frac{1}{2}\gamma\alpha_1^2(\sigma_1^2 - \sigma_2^2) > \frac{1}{2}\gamma\alpha_2^2(\sigma_1^2 - \sigma_2^2).$$

Summing the two inequalities, we get

$$-\mu(e_1) - \frac{1}{2}\gamma(1-\beta_1)^2\lambda(e_1) - ce_1 - \frac{1}{2}\gamma\sigma_1^2\alpha_1^2 + \frac{1}{2}\gamma\alpha_1^2(\sigma_1^2 - \sigma_2^2)$$
$$> -\mu(e_2) - \frac{1}{2}\gamma(1-\beta_2)^2\lambda(e_2) - ce_2 - \frac{1}{2}\gamma\sigma_1^2\alpha_2^2 + \frac{1}{2}\gamma\alpha_2^2(\sigma_1^2 - \sigma_2^2). \qquad (2.32)$$

The expressions on both sides of the inequality simplify to the objective function of the insurer when $\sigma^2 = \sigma_2^2$, which then implies that (α_1, β_1) outperforms the optimal contract (α_2, β_2) when $\sigma^2 = \sigma_2^2$, resulting in a contradiction. We thus conclude that we must have $\alpha_2 \leq \alpha_1$.

Next, we show that $e_1 > e_2$. Again we do this by contradiction. Assume instead that $0 \leq e_1 < e_2$. By the KKT condition for the IC constraints we have:

$$(1-\beta_i)\mu'(e_i) + \frac{1}{2}(1-\beta_i)^2\gamma\lambda'(e_i) + c - \alpha_i = v_i, \qquad (2.33)$$
$$e_i v_i = 0, \quad v_i, e_i \geq 0.$$

As e_2 is strictly positive, we have $v_2 = 0$ in (2.33), leading to

$$(1 - \beta_2)\mu'(e_2) + \frac{1}{2}(1 - \beta_2)^2\gamma\lambda'(e_2) + c - \alpha_2 = 0. \tag{2.34}$$

We can use the above to solve for β_2 as a function of α_2 and e_2 as follows:

$$1 - \beta_2 = \frac{\mu'(e_2) + \sqrt{\mu'(e_2)^2 - 2\gamma(c - \alpha_2)\lambda'(e_2)}}{-\gamma\lambda'(e_2)} := k(e_2, \alpha_2).$$

Therefore, e_2 solves the following optimization problem:

$$\max_{e \geq 0, k(e,\alpha_2) \leq 1} m^o - \mu(e) - \frac{1}{2}\gamma k(e, \alpha_2)^2\lambda(e) - ce - \frac{1}{2}\alpha_2^2\gamma\sigma_2^2.$$

By the optimality of e_2, we conclude that

$$-\mu(e_2) - \frac{1}{2}\gamma k(e_2, \alpha_2)^2\lambda(e_2) - ce_2 \geq -\mu(e_1) - \frac{1}{2}\gamma k(e_1, \alpha_2)^2\lambda(e_1) - ce_1. \tag{2.35}$$

Now consider three cases based on whether α_1, e_1, or both, are non-zero.

1. $\alpha_1 = 0$.

   We know that $\alpha_2 \leq \alpha_1$. Therefore, $\alpha_2 = 0$. It follows from the insurer's optimization problem that $\beta_2 = \beta_1$ and $e_2 = e_1$. This is, however, a contradiction, as we have assumed that $e_1 < e_2$. We therefore must have $\alpha_1 > 0$.

2. $\alpha_1 > 0$ and $e_1 = 0$.

   Take the IC constraint of the agent:

   $$(1 - \beta_1)\mu'(e_1) + \frac{1}{2}(1 - \beta_1)^2\gamma\lambda'(e_1) + c = v_1 + \alpha_1, \tag{2.36}$$

   and $e_1 v_1 = 0$. Since $\alpha_1 > 0$, the insurer can decrease it to $\alpha_1 = 0$, and instead increasing v_1; increasing v_1 is possible as when $e_1 = 0$, v_1 can be strictly positive. Decreasing α_1 in this way increases the insurer's payoff without affecting β_1, e_1, contradicting the optimality of the contract. Therefore, this case is also impossible.

3. $\alpha_1 > 0$ and $e_1 > 0$.

   From the IC constraint at $\sigma^2 = \sigma_1^2$, we have:

   $$(1 - \beta_1)\mu'(e_1) + \frac{1}{2}(1 - \beta_1)^2\gamma\lambda'(e_1) + c - \alpha_1 = 0. \tag{2.37}$$

   From this, we find β_1 as a function of α_1 and e_1,

   $$1 - \beta_1 = \frac{\mu'(e_1) + \sqrt{\mu'(e_1)^2 - 2\gamma(c - \alpha_1)\lambda'(e_1)}}{-\gamma\lambda'(e_1)}.$$

Therefore, e_1 solves the following optimization problem:

$$\max_{e \geq 0, k(e, \alpha_1) \leq 1} h(e) := \quad m^o - \mu(e) - \frac{1}{2}\gamma k(e, \alpha_1)^2 \lambda(e) - ce - \frac{1}{2}\sigma_1^2 \gamma \alpha_1^2. \quad (2.38)$$

Rewrite $h(e_1)$ as follows:

$$
\begin{aligned}
h(e_1) \quad = \quad & m^o - \mu(e_1) - \frac{1}{2}\gamma k(e_1, \alpha_2)^2 \lambda(e_1) - ce_1 \\
& - \frac{1}{2}\sigma_1^2 \gamma \alpha_1^2 + \frac{1}{2}\gamma(k(e_1, \alpha_2)^2 - k(e_1, \alpha_1)^2)\lambda(e_1). \quad (2.39)
\end{aligned}
$$

Then $h(e_2) - h(e_1)$ is given by

$$
\begin{aligned}
& h(e_2) - h(e_1) \\
= \quad & (-\mu(e_2) - \frac{1}{2}\gamma k(e_2, \alpha_2)^2 \lambda(e_2) - ce_2) - (-\mu(e_1) - \frac{1}{2}\gamma k(e_1, \alpha_2)^2 \lambda(e_1) - ce_1) \\
& + \frac{1}{2}\gamma[(k(e_2, \alpha_2)^2 - k(e_2, \alpha_1)^2)\lambda(e_2) - (k(e_1, \alpha_2)^2 - k(e_1, \alpha_1)^2)\lambda(e_1)]. \quad (2.40)
\end{aligned}
$$

First, note that (2.40) is non-negative by (2.35). Next, take $0 \leq \alpha_2 \leq \alpha_1 \leq c$ ($\alpha_2 \leq \alpha_1$ follows from the proof at the beginning of this theorem, and the lower and upper bounds follow from Lemma 2.4). Assuming $(k(e, \alpha_2)^2 - k(e, \alpha_1)^2)\lambda(e)$ is non-decreasing, it follows (2.40) is also non-negative. Therefore, $h(e_2) \geq h(e_1)$, which is a contradiction as e_1 is the maximizer of $h(e)$ in the optimization problem (2.38). We thus conclude $e_1 \geq e_2$.

For the last part of the theorem, we compare the presence of any pre-screening quality $\sigma = \sigma_1$ to the case of $\sigma = \infty$ (uninformative pre-screening). We want to show that when $k(e, 0)^2 \lambda(e) - k(e, \alpha)^2 \lambda(e)$ is non-decreasing in e, we have $e_\infty \leq e_1$. Note that when σ is infinity, the optimal discount factor is zero. As a result, the proof follows that of the first part of the theorem with $\alpha_2 = 0$.

CHAPTER 3

Insuring Clients with Dependent Risks

The previous chapter presented two basic contract models, one over a single period, the other over two periods. In both cases only a single agent is involved. Using these models we discussed what it means to premium-discriminate, what it means to pre-screen or post-screen, and how moral hazard is manifested in the agent's utility, the insurer's utility, and the agent's state of security. In particular, we saw how pre-screening can be an effective tool for the insurer to assess the insured's risk and mitigate moral hazard.

In this chapter we will take the single-insurer single-agent model and extend it to the scenario of a single insurer and multiple agents with dependent risks over a single period. For much of this chapter we will focus on the case of two interdependent agents without loss of generality. In particular, we analyze the impact of risk interdependency and pre-screening on the optimal contract and the agents' effort, when they are risk neutral and risk averse, respectively; the purpose of investigating the former is so that we can distinguish the effect of risk aversion from that of interdependence.

The motivation for considering agents with dependent risks was discussed briefly in Section 1.2 as one of the unique features of cyber insurance. A major cause of this is business dependencies between organizations as a result of outsourcing or supply chain relationships. In these cases the state of security of one firm depends not only on its own effort but also that of other firms [47, 48, 56, 64, 65, 75, 77, 80, 89, 109]. With increasing connectedness among today's businesses, risks can spill over easily and quickly from one firm to another. For instance, a breach at a credit card processing vendor can lead to major losses by retailers, or an outage at a network service provider (such as Amazon or Microsoft cloud services) can result in business interruption to a large number of customers. If both the provider and its (one or more) customers happen to be underwritten by the same insurer, then managing this type of system risk across a policy portfolio becomes vitally important.

More broadly, systemic risk typically means correlated or aggregate risk across a portfolio of policies that can result in catastrophic losses [49]. It could arise from common technology dependencies among policy holders rather than dependencies of each other. A prime example of this is when a common vulnerability or system configuration shared across many policy holders is exploited, leading to multiple, simultaneous breaches and subsequent insurance claims. A case in point is the massive WannaCry and NotPetya attacks of 2016, and more recently the

SolarWinds hack of 2020, all caused by exploiting a common technology dependency and any common vulnerability therein across many firms [113].

This chapter will primarily focus on systemic risks caused by dependent relationship among policy holders themselves rather than their common dependency on an outside party. However, the qualitative conclusions we reach are equally applicable in the latter case. As we will see, this type of risk dependency can lead firms to free ride on other firms' efforts and under-invest in security [76, 82, 127], a phenomenon that becomes relevant in our analysis. The model presented in this chapter assumes certain quantitative knowledge of the underlying risk dependency. In practice, an insurer can measure this dependency by collecting vendor/supplier information from an insured, often followed by Monte Carlo style of scenario modeling and simulation.

3.1 A MODEL OF TWO AGENTS

Consider two agents both in the market for cyber insurance. Their risks are interdependent, in that the effort exerted by one agent affects not only himself, but also the loss experienced by the other. This is modeled as follows:

$$L_{e_1,e_2}^{(i)} \sim \mathcal{N}(\mu(e_i + x \cdot e_{-i}), \lambda(e_i + x \cdot e_{-i})). \tag{3.1}$$

Here we have used the convention $\{-i\} = \{1,2\} - \{i\}$, and $L_{e_1,e_2}^{(i)}$ is a random variable denoting the loss that agent i experiences, given both agents' efforts. There is an *interdependence factor*, denoted by $x \in [0, 1)$, that captures the level of mutual reliance between the two agents.[1] Note that this is not a unique modeling choice and is indeed a simplification. A more general way of expressing correlated risks would be to model the losses as jointly distributed; more on this is discussed in Section 3.4.

We assume the agents' utilities are again given by (2.5) and (2.3) for risk-neutral and risk-averse agents, respectively, with the loss distributions replaced by the above expression. We allow the two agents to have different effort cost c_1, c_2, as well as different risk attitudes γ_1, γ_2.

The insurer can again conduct a pre-screening assessment, $S_{e_i} = e_i + W_i$, on each agent i, where W_i is a zero-mean Gaussian noise with variance σ_i^2. We assume that W_1 and W_2 are independent,[2] and that $S_{e_1}, S_{e_2}, L_{e_1,e_2}^{(1)}, L_{e_1,e_2}^{(2)}$ are conditionally independent given e_1, e_2. The contract is again in the form of a triple (b, α, β): base premium, premium discount, and fraction of coverage.

Similar to the single-agent case, we need to evaluate the agents' options outside a contract. These will then be used to impose the individual rationality constraints in determining the terms of the contracts. However, compared to the single-agent case, the outside option of one agent is

---

[1]A single parameter x for both i and $-i$ suggests a symmetric dependence; this can be generalized to an asymmetric setting with $x_{ij} \neq x_{ji}$.

[2]An example and discussion on correlated pre-screening noises can be found in Section 3.4.

now influenced by the participation choice of the other agent. Specifically, we need to evaluate the agents' utilities as well as potential contracts in the following three scenarios.

(i) Neither agent enters a contract; their dependent relationship leads them to best respond to each other outside insurance.

(ii) One enters a contract, while the other opts out; the insurer becomes part of the strategic interaction through her contract terms with the agent who opts in.

(iii) Both agents purchase contracts; in this case the insurer jointly determines the contract terms for each agent.

Here, Case (ii) is the outside option for agents in Case (iii), and Case (i) is the outside option for agents in Case (ii). In order to evaluate the participation constraints of agents when both purchase insurance contracts (Case (iii)), we first need to find the optimal contracts and agents' utilities in Cases (i) and (ii). We therefore proceed by evaluating the agents' utilities in each case, followed by solving the insurer's contract design problem. We do this in Sections 3.2 and 3.3 for risk-neutral and risk-averse agents, respectively.

A note on notations used in this chapter.   Because of the three games we need to consider, we will use superscripts "oo", "io", and "ii" to denote each of these three cases when parameters and quantities are specific to one case: "o" stands for opting out (of a contract), "i" stands for opting in, and their position in the superscript signifies the first vs. the second agent. These are not to be confused with the subscripts "i" and "$-i$" which denote one or the other agent i, $i = 1, 2$. In addition, we will continue to use U and V to denote the agents' and the insurer's utilities, $\overline{U}$ and $\overline{V}$ their expectations, and u and v their optimal values, respectively.

Tables summarizing the notations used in this chapter can be found in Section 3.7.

3.2   TWO RISK-NEUTRAL AGENTS

The intention of studying risk-neutral agents is so we can set aside the effect of risk aversion and solely focus on the effect of risk dependence.

In order to evaluate the agents' opt-out options and find the optimal contract, the insurer's problem and the agents' utilities need to be studied under the three cases listed above. Below we analyze these three cases and discuss the role of pre-screening and the contracts' effect on network security.

3.2.1   CASE (I): NEITHER AGENT ENTERS A CONTRACT

This case is illustrated in Figure 3.1. Let G^{oo} denote the game between the two risk-neutral agents who have both decided not to purchase cyber insurance contracts. In this game, the agents' efforts e_1, e_2 are their actions, and their expected payoffs with unit costs of effort $c_1, c_2 > 0$ are

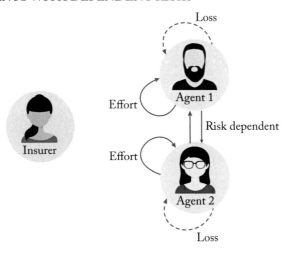

Figure 3.1: Neither agent purchases insurance, forming a game outside contract.

given by:

$$\overline{U}_i^{oo}(e_1, e_2) = -\mu(e_i + xe_{-i}) - c_i e_i, \tag{3.2}$$

where $i \in \{1, 2\}$. The best response, denoted by B^{oo}, of each agent is therefore given by

$$B_i^{oo}(e_{-i}) = \arg\max_{e_i \geq 0} \; -\mu(e_i + xe_{-i}) - c_i e_i. \tag{3.3}$$

The above optimization problem is convex, and has the following solution:

$$\begin{aligned}
B_i^{oo}(e_{-i}) &= (m_i - xe_{-i})^+, \quad \text{where} \\
m_i &= \arg\min_{e \geq 0} \mu(e) + c_i e, \quad i = 1, 2,
\end{aligned} \tag{3.4}$$

and $(a)^+ = \max\{a, 0\}$; details can be found in Appendix 3.8. The Nash equilibrium (NE) is given by the fixed point of the best-response mappings $B_1^{oo}(e_2)$ and $B_2^{oo}(e_1)$:

$$e_1 = (m_1 - xe_2)^+ \quad \text{and} \quad e_2 = (m_2 - xe_1)^+. \tag{3.5}$$

In Appendix 3.8 we show that given $0 \leq x < 1$, the system of equations (3.5) has a unique fixed point, i.e., the unique NE, denoted by $e_i^*(m_i, m_{-i})$:

$$e_i^*(m_i, m_{-i}) = \begin{cases} \frac{m_i - x \cdot m_{-i}}{1 - x^2} & \text{if } m_i \geq x \cdot m_{-i} \text{ and } m_{-i} \geq x \cdot m_i \\ 0 & \text{if } m_i \leq x \cdot m_{-i} \\ m_i & \text{if } m_{-i} \leq x \cdot m_i \end{cases}. \tag{3.6}$$

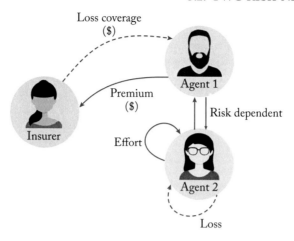

Figure 3.2: One agent purchases insurance, forming a game that now involves the insurer.

We will also use e_i^{oo} to denote these equilibrium efforts outside the contract. Accordingly, $u_i^{oo} := \overline{U}_i^{oo}(e_1^*(m_1, m_2), e_2^*(m_2, m_1))$ is the utility of agent i in the equilibrium when neither agent chooses to enter a contract. As we will see shortly, an insurer uses her knowledge of u_i^{oo} to evaluate agents' outside options when proposing optimal contracts.

In parallel to G^{oo}, where two agents best responds in the absence of insurance, we will also consider the notion of *socially optimal* levels of effort, denoted by $\tilde{e}_i$, which is the solution to the problem $\max_{e_i} \sum_{i=1,2} \overline{U}_i^{oo}(e_i, e_{-i})$. This will be used later as a benchmark for comparison.[3]

3.2.2   CASE (II): ONE AND ONLY ONE ENTERS A CONTRACT

This case is illustrated in Figure 3.2. Assume without loss of generality that agent 1 enters a contract, while agent 2 opts out. We use G^{io} to denote the game between the insured agent 1 and uninsured agent 2. The agents' expected payoff in this case is:

$$\overline{U}_1^{io}(e_1, e_2, b_1, \alpha_1, \beta_1) = -b_1 - (c_1 - \alpha_1)e_1 - (1 - \beta_1)\mu(e_1 + xe_2), \qquad (3.7)$$

$$\overline{U}_2^{io}(e_1, e_2) = -\mu(e_2 + xe_1) - c_2e_2. \qquad (3.8)$$

Let $B_1^{io}(e_2) = \arg\max_{e_1 \geq 0} \overline{U}_1^{io}(e_1, e_2, b_1, \alpha_1, \beta_1)$ denote the best response of agent 1. This optimization problem is convex and has a solution given by:

$$B_1^{io}(e_2) = (m_1(\alpha_1, \beta_1) - xe_2)^+, \text{ where} \qquad (3.9)$$

$$m_1(\alpha_1, \beta_1) = \arg\min_{e \geq 0}\{(c_1 - \alpha_1)e + (1 - \beta_1)\mu(e)\}. \qquad (3.10)$$

[3]The equilibrium in G^{oo} is sometimes called anarchy; $\tilde{e}_i$ may be obtained using a social planner.

For the uninsured agent 2, it is easy to see that the best-response function is given by $B_2^{oo}(e_1)$, the same best-response function in game G^{oo}. We can now find the NE as the fixed point of the best-response mappings. Agents' efforts at the equilibrium are $e_1^*(m_1(\alpha_1, \beta_1), m_2)$ and $e_2^*(m_2, m_1(\alpha_1, \beta_1))$, as defined in (3.6).

Let $\overline{V}^{io}(b_1, \alpha_1, \beta_1, e_1, e_2)$ denote the insurer's utility, when agent 2 opts out, the insurer offers contract (b_1, α_1, β_1) to agent 1, and agents exert efforts e_1, e_2, respectively. The optimal contract offered by the insurer to the participating agent is the solution to the following problem:

$$\max_{b_1, \alpha_1, 0 \leq \beta_1 \leq 1, e_1^*, e_2^*} \overline{V}^{io}(b_1, \alpha_1, \beta_1, e_1^*, e_2^*) = b_1 - \alpha_1 e_1^* - \beta_1 \mu(e_1^* + x e_2^*) \quad (3.11)$$

$$\text{s.t.} \quad \text{(IR)} \quad \overline{U}_1^{io}(e_1^*, e_2^*, b_1, \alpha_1, \beta_1) \geq u_1^{oo}$$

$$\text{(IC)} \quad e_1^*, e_2^* \text{ are the NE of game } G^{io}.$$

Similar to a result presented in Chapter 2, the IR constraint is binding under the optimal contract. Therefore, we can re-write the insurer's problem by replacing the base premium b_1:

$$\max_{\alpha_1, 0 \leq \beta_1 \leq 1, e_1^*, e_2^*} -u_1^{oo} - \mu(e_1^* + x e_2^*) - c_1 e_1^* \quad (3.12)$$

$$\text{s.t.} \quad \text{(IC)} \quad e_1^*, e_2^* \text{ are the NE of game } G^{io}. \quad (3.13)$$

Let u_2^{io} be the second agent's utility when the insurer offers the *optimal contract* to the first agent and the second agent opts out. The insurer can calculate u_2^{io} by finding the optimal contract in problem (3.12) and the resulting NE of game G^{io}.

Similarly, let u_1^{oi} denote the first agent's utility when he opts out and the second agent purchases the *optimal contract*. The insurer uses her knowledge of u_2^{io} and u_1^{oi} to design a pair of contracts to attract both agents as we discuss below.

3.2.3   CASE (III): BOTH AGENTS PURCHASE A CONTRACT

This case is illustrated in Figure 3.3. Let G^{ii} denote the game between the two agents when they are both in a contract. Assume the insurer offers each agent i a contract (b_i, α_i, β_i). The expected utility of the agents when both purchase contracts is given by

$$\overline{U}_i^{ii}(e_1, e_2, b_i, \alpha_i, \beta_i) = -b_i - (c_i - \alpha_i)e_i - (1 - \beta_i)\mu(e_i + x e_{-i}). \quad (3.14)$$

Following steps similar to those in the previous case, B_i^{ii}, the best-response function of agent i, is given by

$$B_i^{ii}(e_{-i}) = (m_i(\alpha_i, \beta_i) - x e_{-i})^+, \quad (3.15)$$

where $m_i(\alpha_i, \beta_i)$ is the solution to

$$m_i(\alpha_i, \beta_i) = \arg\min_{e \geq 0}\{(c_i - \alpha_i)e + (1 - \beta_i)\mu(e)\}. \quad (3.16)$$

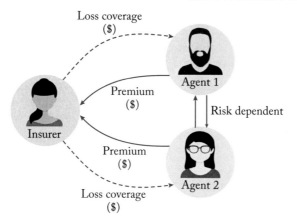

Figure 3.3: Both agents purchase insurance, allowing the insurer to determine the two contract terms jointly.

The agents' efforts at the Nash equilibrium are again the fixed point of the best-response mappings, and is given by $e_i^*(m_i(\alpha_i, \beta_i), m_{-i}(\alpha_{-i}, \beta_{-i}))$, with $e_i^*(.,.)$ defined in (3.6).

To write the insurer's problem, note that the outside option of agent 1 (resp. 2) in this game is his utility in the game G^{oi} (resp. G^{io}). It follows that the optimal contracts offered by the insurer to the agents are the solution to the following optimization problem:

$$\max_{b_i,\alpha_i,\beta_i,e_i^*,i=1,2} \sum_{i=1,2} b_i - \alpha_i e_i^* - \beta_i \mu(e_i^* + x e_{-i}^*) \qquad (3.17)$$

$$\text{s.t.} \quad (\text{IR}) \quad \overline{U}_1^{ii}(e_1^*, e_2^*, b_1, \alpha_1, \beta_1) \geq u_1^{oi},$$

$$\overline{U}_2^{ii}(e_1^*, e_2^*, b_2, \alpha_2, \beta_2) \geq u_2^{io},$$

$$(\text{IC}) \quad e_1^*, e_2^* \quad \text{are the NE of game } G^{ii}.$$

The IR constraints can again be shown to be binding. Therefore, the insurer's contract design problem for two risk-neutral agents is given by:

$$v^{ii} := \max_{\alpha_1,\beta_1,\alpha_2,\beta_2,e_1^*,e_2^*} -u_1^{oi} - u_2^{io} - \sum_{i=1,2} (\mu(e_i^* + x e_{-i}^*) - c_i e_i^*) \qquad (3.18)$$

$$\text{s.t.} \quad e_1^*, e_2^* \quad \text{are the NE of game } G^{ii}.$$

The optimal efforts (along with the optimal contract parameter values) that maximize the above objective function will be denoted by e_i^{ii}.

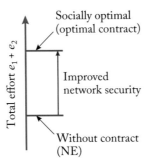

Figure 3.4: The effort gap outside and inside contracts.

3.2.4   OPTIMAL CONTRACTS FOR TWO RISK-NEUTRAL AGENTS

We now present the properties of the contracts derived from the optimization problem (3.18) and their impact on the agents' efforts. Recall the following notation: e_i^{oo} denotes the effort of agent i when insurance is not available; e_i^{ii} denotes the effort of agent i in the solution to (3.18), i.e., when purchasing the optimal contract; $\tilde{e}_i$ denotes the effort level of agent i in the socially optimal outcome, i.e, the efforts maximizing the sum of agents' utilities.

Theorem 3.1   *A profit-maximizing insurer offers contracts to both agents, with the following properties.*

1. *$e_i^{ii} = \tilde{e}_i, i = 1, 2$. That is, the agents exert socially optimal effort levels in the optimal contract.*

2. *$e_1^{ii} + e_2^{ii} \geq e_1^{oo} + e_2^{oo}$. That is, when both agents purchase the optimal insurance contracts, the overall effort exerted toward security increases compared to the no-insurance scenario.*

3. *$v^{ii} \geq \overline{U}_1(\tilde{e}_1, \tilde{e}_2) + \overline{U}_2(\tilde{e}_1, \tilde{e}_2) - \overline{U}_1(e_1^{oo}, e_2^{oo}) - \overline{U}_2(e_1^{oo}, e_2^{oo})$. That is, the insurer's profit is higher than the gap between agents' welfare at the socially optimal solution and the no-insurance equilibrium.*

Theorem 3.1 is illustrated in Figure 3.4 and interpreted as follows.

First, recall that, as discussed in Section 2.1.4, the insurer cannot make profit from offering contracts to a single risk-neutral agent, as there is no risk transfer from the agent to an insurer. However, we now observe the opposite: the insurer can make profit when offering contracts to *two* interdependent risk-neutral agents. Clearly, this improvement stems from the agents' interdependency.

Specifically, due to interdependency, agents under-invest in security at the no-insurance equilibrium. This leads to a profit opportunity for the insurer, in which she uses her (accurate) pre-screening assessments to offer premium discounts and in turn requires the agents to exert higher efforts. This increase in efforts is in the insurer's interest, as it lowers the risks of both of

Figure 3.5: The profit gap outside and inside contracts.

her contracts. This effect can be viewed as the insurer "selling commitment" to the agents. That is, the insurer is effectively providing each agent with the commitment of the other agent to also exert higher effort if he commits to exerting high effort.

Indeed, in this case the insurer offers full coverage of losses, and in turn the agents exert socially optimal levels of effort. To see this, we note that the insurer's contract design problem in (3.18) does not contain α_i and β_i in the objective function. Therefore, the insurer can choose to set $\alpha_i = c_i$ and $\beta_i = 1$ (i.e., full reimbursement of the agent's effort and full coverage of loss). Also note the constraint on the agents' effort is given by, for $i = 1, 2$:

$$m_i(\alpha_i, \beta_i) = \arg\min_{e \geq 0}(1 - \beta_i)\mu(e) + (c_i - \alpha_i)e$$
$$e_i = e_i^*(m_i(\alpha_i, \beta_i), m_{-i}(\alpha_{-i}, \beta_{-i})).$$

Therefore, any non-negative effort will suffice. Consequently, the insurer's problem simplifies to

$$\max_{e_1, e_2 \geq 0} -u_1^{oi} - u_2^{io} - \mu(e_1 + xe_2) - \mu(e_2 + xe_1) - c_1e_1 - c_2e_2,$$

which is equivalent to maximizing $\overline{U}_1(e_1, e_2) + \overline{U}_2(e_1, e_2)$ with constraints $e_1, e_2 \geq 0$, meaning that the socially optimal strategies also maximize the insurer's profit in the optimal contract. In other words, $e_i^{ii} = \tilde{e}_i$, $i = 1, 2$. This is also essentially how we prove part 1 of Theorem 3.1. Proof of parts 2 and 3 can be found in Appendix 3.8.

Second, part 3 of the theorem shows that the profit opportunity for the insurer is even higher than the welfare gap between the socially optimal and Nash equilibrium outcomes, as shown in Figure 3.5. This is due to the fact that the outside option from the contract for agent i is an outcome in which the insurer offers a contract (only) to agent $-i$. The insurer will select this contract in a way that it requires agent $-i$ to exert low effort and get high coverage, effectively forcing agent i to bear the full cost of effort, leading to a utility lower than the no-insurance Nash equilibrium for agent i. As the agents' IR constraints are also binding, it follows that the insurer's profit is in fact the gap between the welfare attained under the optimal contract and the welfare at these low-payoff, unilateral opt-out outcomes.

Finally, note that the statements of this theorem do not depend on the pre-screening noises $\sigma_i < \infty$. This is because the expected utilities and consequent effort choices of risk-neutral agents are only sensitive to the mean, but not the variances of uncertainties in the problem parameters. As a result, under the assumption of zero-mean noise in the pre-screening assessments, the agents' behavior are independent of σ.

3.3 TWO RISK-AVERSE AGENTS

We next analyze the case of two risk-averse agents. Again, as discussed in Section 3.1, in order to evaluate the agents' individual rationality constraints and find the optimal contracts, we need to account for three possible cases based on the agents' participation decisions. The ensuing analysis is similar to that presented in Section 3.2, by replacing the agent's utility functions with their risk-averse versions and solving the resulting optimization problems. The details can be found in [71, 73], and are not reproduced here for brevity.

Following similar analysis, the insurer's optimization problem can be simplified as follows:

$$v^{ii} = \max_{\alpha_i, \beta_i, e_i^*, i=1,2} \ln(-u_1^{oi})/\gamma_1 + \ln(-u_2^{io})/\gamma_2 -$$

$$\sum_{i=1,2} \left(\mu(e_i^* + xe_{-i}^*) + \frac{\gamma_i}{2}(1-\beta_i)^2 \lambda(e_i^* + xe_{-i}^*) + c_i e_i^* + \frac{\gamma_i}{2}\alpha_i^2 \sigma_i^2 \right)$$

$$\text{s.t.} \quad e_1^*, e_2^* \text{ are the NE of game } G^{ii}. \tag{3.19}$$

Below we summarize how different problem parameters, particularly the availability of pre-screening, affect the insurer's profit in the optimal contracts, as well as the system's state of security. We first consider the utility of the insurer. Note that the insurer always has the option to not use the outcome of pre-screening by setting $\alpha = 0$ in the contract. Therefore, the insurer's utility in the optimal contract with pre-screening is no less than that in the optimal contract without pre-screening; i.e., the availability of pre-screening is in the insurer's interest.

We next turn to the effect of pre-screening on the state of network security, measured by the total effort toward security.

Theorem 3.2 *Let $m_i = \arg\min_{e \geq 0} \mu(e) + \frac{1}{2}\gamma_i \lambda(e) + c_i e$ and let e_i^{ii} and e_i^{oo} denote the effort of agent i in the optimal contract and in the no-insurance equilibrium, respectively.*

1. *If pre-screening is perfect, i.e., $\sigma_1 = \sigma_2 = 0$, then $e_1^{ii} + e_2^{ii} \geq e_1^{oo} + e_2^{oo}$ if for $i = 1, 2$ we have:*

$$(i) \quad \mu'(m_i) < \frac{-c_i + xc_{-i}}{1 - x^2}, \tag{3.20}$$

$$(ii) \quad (\mu')^{-1}\left(\frac{-c_i + xc_{-i}}{1 - x^2}\right) \geq x(\mu')^{-1}\left(\frac{-c_{-i} + xc_i}{1 - x^2}\right). \tag{3.21}$$

That is, under these conditions, insurance improves network security compared to the no-insurance scenario.

2. *If pre-screening is uninformative, i.e., $\sigma_1 = \sigma_2 = \infty$, then $e_1^{ii} + e_2^{ii} \leq e_1^{oo} + e_2^{oo}$. That is, the insurance contract without pre-screening worsens network security as compared to the no-insurance scenario.*

The results of Theorem 3.2 can be intuitively interpreted as follows. By Theorem 2.1, with a single risk-averse agent, the insurer profits from the agent's interest in risk transfer. However, the introduction of insurance *always* reduces network security. In contrast, Theorem 3.2 shows that with interdependent agents, the network security can actually improve, and equally importantly, the insurer continues to make a profit.

To see how the agents' interdependency plays a role in the improvement of security, note that the insurer uses pre-screening and offers premium discounts accordingly in order to incentivize the interdependent agents to increase their effort levels. Providing such incentives is in the insurer's interest, as higher effort exerted by the agent decreases both agents' risk, and consequently, the coverage required by the insurer once losses are realized. Note also that it is the availability of (accurate) pre-screening that provides the required tools for the insurer in designing such incentives; otherwise, as shown in part (2) of the theorem, improving network security is impossible.

The conditions of part (1) of the theorem can also be interpreted as follows. The first condition imposes a restriction on the derivative of μ, so that the decrease in loss as a function of effort is faster than the normalized cost of effort; as a result, the insurer will have the option to make more profit through loss reduction (by encouraging agents to exert higher effort). The second condition imposes a restriction on the agents' cost of effort and guarantees that both agents exert positive effort. Specifically, when the two agents' effort costs are sufficiently similar, this condition is satisfied, and both agents exert non-zero effort.

3.4   MULTIPLE AGENTS, CORRELATED LOSSES, AND THE INSURER'S RISK AVERSION

Results presented in the preceding section can be extended in various directions, and the key observation remains valid that insurance can effectively incentivize higher effort when equipped with sufficiently accurate pre-screening capabilities. We summarize some of these below; more in-depth development can be found in [73].

N **homogeneous risk-averse agents.**   Consider N homogeneous risk-averse agents given by $\gamma_i = \gamma$, $c_i = c$, and $\sigma_i = \sigma$, $\forall i$. In this case, we can obtain a result similar to Theorem 3.2. In particular, under perfect pre-screening ($\sigma = 0$), the sufficient and necessary condition for the network security to improve after the introduction of insurance is $\mu'(m) < -\frac{c}{1+(N-1)x}$, which, when setting $N = 2$, reduces to the condition in Theorem 3.2 with $c_i = c$. Furthermore, with sufficiently accurate yet imperfect pre-screening, insurance can again lead to improvement in the state of security compared to the no-insurance equilibrium.

Correlated losses and the insurer's risk aversion.   Continuing with the setting of N homogeneous agents, now consider the case where the losses of these agents are not only distributionally dependent but also correlated in their realizations defined as follows. For simplicity, we will assume that pre-screening is perfect ($\sigma = 0$ and $S_i = e_i$). Let θ be the covariance between any two losses, that is,

$$Cov(L_e^i, L_e^j) = \theta, \quad \forall i \neq j, \tag{3.22}$$

where the vector $(L_e^1, \cdots, L_e^N)$ is given by a multivariate Gaussian distribution.

Suppose the insurer is also risk averse, with risk attitude $\delta \geq 0$. We can show that the sufficient and necessary condition for agents to exert higher effort inside than outside the contract is $\mu'(m) + \frac{1}{2}\frac{\delta\gamma}{\gamma+\delta}\lambda'(m) + \frac{c}{1+(N-1)x} < 0$, for a similarly defined $m = \arg\min_{e \geq 0} \mu(e) + \frac{\gamma}{2}\lambda(e) + c$.

It is easy to see that this condition reduces to that shown in the preceding N-homogeneous-agent case if we set $\delta = 0$. This condition is more likely to be satisfied for larger values of δ, i.e., a more risk-averse insurer. In other words, if the insurer is more risk averse, it is more likely that she encourages agents to exert higher effort as compared to their efforts outside of the contract. In the extreme case, when $\delta = \infty$, the condition is always satisfied.

At the same time, we can also show that if the agents' losses are more correlated, a risk-averse insurer encourages the agents to exert more effort. This is because with correlated losses, it is more likely for losses to happen simultaneously as compared to a scenario with independent losses. Note that when $\delta = 0$, the problem becomes independent of θ, meaning that the covariance between any two losses does not affect the optimal contract or the agents' efforts if the insurer is risk neutral.

3.5   NUMERICAL RESULTS

We present some numerical examples of the results presented in this chapter. Our main focus is on the impact of pre-screening noise in various scenarios. Throughout the first part of this section we use the following parameters:

$$\mu(e) = \frac{10}{e+1}, \quad \lambda(e) = \frac{10}{(e+1)^2}, \quad c = 2, \quad \gamma = 1. \tag{3.23}$$

3.5.1   IMPACT OF PRE-SCREENING NOISE

Two homogeneous risk-averse agents.   Consider two homogeneous agents with interdependence factor $x = 0.5$. Figure 3.6 shows the insurer's utility as a function of the quality of pre-screening: the insurer's profit decreases as the pre-screening accuracy decreases. Figure 3.7 shows network security as a function of pre-screening noise. Here, the conditions of Theorem 3.2 are satisfied. As we can see, security under the contract is higher than that without insurance for small values of σ; but as σ increases, security worsens and drops below that without contract.

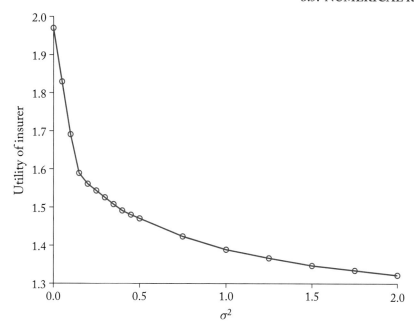

Figure 3.6: Insurer utility vs. σ^2; two homogeneous risk-averse agents.

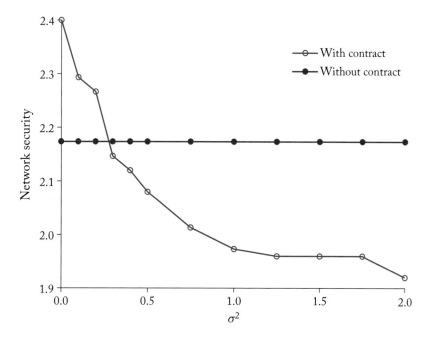

Figure 3.7: Network security ($e_1 + e_2$) vs. σ^2; two homogeneous risk-averse agents.

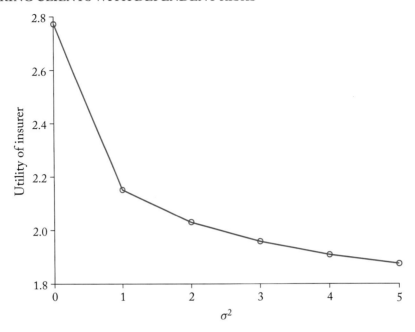

Figure 3.8: Insurer profit vs. σ^2 with two heterogeneous risk-averse agents.

Two heterogeneous risk-averse agents. We next consider two heterogeneous agents with the following parameters:

$$\mu(e) = \frac{10}{e+1}, \quad \lambda(e) = \frac{10}{(e+1)^2}, \quad c_1 = 1, \quad c_2 = 1.1$$

$$\gamma_1 = 1.2 \quad \gamma_2 = 1, \quad x = 0.5. \tag{3.24}$$

We assume the pre-screening noise (σ^2) is the same for both agents. These parameters together satisfy the condition of Theorem 3.2. Again, Figure 3.8 shows that the insurer's profit decreases as pre-screening becomes less accurate, and Figure 3.9 shows that the introduction of insurance can indeed improve the state of network security provided the pre-screening is sufficiently accurate.

3.5.2 THE SUFFICIENT CONDITIONS OF THEOREM 3.2

Consider an example with parameters similar to those given in (3.24), except that $\gamma_1 = 1.5$ and $c_2 = 1.5$. In this case, it can be verified that the conditions of Theorem 3.2 do not hold. However, Figure 3.10 shows that network security improves after the introduction of insurance. This example shows that the sufficient conditions in Theorem 3.2 are not necessary.

Consider again the same parameters given in (3.24), except $x = 0.15$. In this case, it can again be verified that the conditions of Theorem 3.2 do not hold. Figure 3.11 shows that the

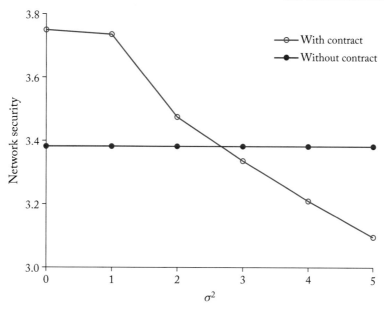

Figure 3.9: Network security $(e_1 + e_2)$ vs. σ^2 with two heterogeneous risk-averse agents.

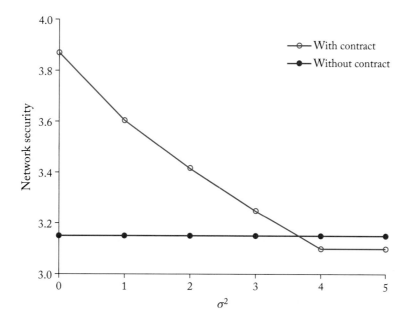

Figure 3.10: Network security $(e_1 + e_2)$ vs. σ^2 with two heterogeneous risk-averse agents. In this example, the conditions of Theorem 3.2 do not hold but network security improves after the introduction of insurance.

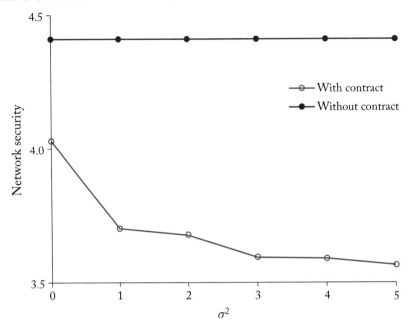

Figure 3.11: Network security ($e_1 + e_2$) vs. σ^2 with two heterogeneous risk-averse agents. In this example, the conditions of Theorem 3.2 do not hold, and network security worsens after the introduction of insurance.

network security worsens with the introduction of insurance and thus the sufficient conditions are meaningful.

3.5.3   LOSS AND PRE-SCREENING NOISE DISTRIBUTIONS

Throughout our analysis, we have assumed that losses and pre-screening outcomes are normally distributed. Below we provide a numerical example under the assumption of exponentially distributed losses and uniformly distributed pre-screening outcomes. We illustrate how our previous observations hold in this instance as well. Consider two risk-averse agents with the following parameters:

$$\gamma_1 = \gamma_2 = 0.9, \; c_1 = 0.25, \; c_2 = 0.5, x = 0.5$$
$$E(L^i_{e_1,e_2}) = \mu(e_i + xe_{-i}) = \frac{1}{1 + e_i + xe_{-i}}$$
$$L^i_{e_1,e_2} \sim \exp(\frac{1}{\mu(e_i + xe_{-i})}),$$
$$S_{e_i} = e_i + W_i, \; W_i \sim \text{Unif}(-w, w), \; i = 1, 2,$$

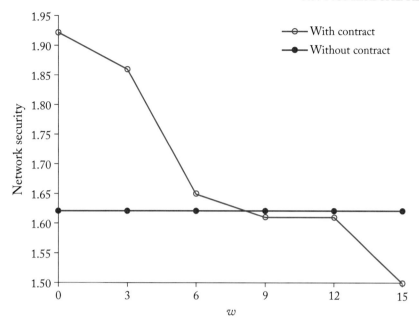

Figure 3.12: Network security $(e_1 + e_2)$ vs. pre-screening noise with two heterogeneous risk-averse agents with exponentially distributed interdependent losses.

where W_1, W_2 are independent and uniformly distributed in interval $[-w, w]$. Figure 3.12 shows network security in this example. Again, we see that when pre-screening is sufficiently accurate (w small), by exploiting agents' interdependence, the insurer can design contracts in a way that network security inside the contract is higher than prior to the introduction of insurance. In contrast, when pre-screening is not accurate enough (w large), network security inside the contract falls bellow network security outside the contract.

3.5.4   CORRELATED PRE-SCREENING NOISES

Throughout our analysis, we have also assumed that pre-screening noises W_1 and W_2 are independent. The correlation between W_1 and W_2 does not affect our results when the insurer is risk neutral. As noted earlier, this is because a risk-neutral insurer is not sensitive to the variance and covariance of the pre-screening outcomes. On the other hand, correlation between the two can impact the utility of a risk-averse insurer in much the same way that loss covariance does (see Section 3.4). The following Figure 3.13 illustrates network security as a function of pre-screening noise correlation. The parameters of this example are as follows: $\mu(e) = \frac{100}{2e+1}$, $\lambda(e) = \frac{100}{(2e+1)^2}$, $c_1 = c_2 = 1$, $\gamma_1 = \gamma_2 = 0.5$, $\delta = 0.1$, $\sigma^2 = 2$, $x = 0.5$.

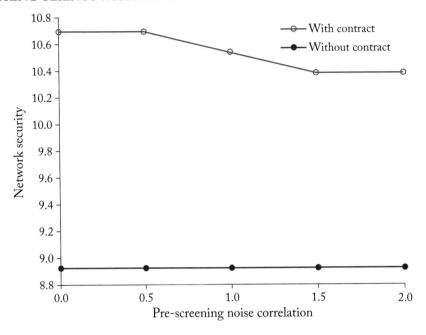

Figure 3.13: Network security as a function of pre-screening noise correlation.

3.6   CHAPTER SUMMARY

This chapter extends the previous single-insurer single-agent model to a single-insurer multi-agent model where the multiple agents' risks are interdependent. Analysis of this model shows that whereas in the case of a single agent his effort is always lower within an insurance contract than without, the presence of two risk-dependent agents gives rise to a unique "profit gap," or a win-win opportunity for the insurer to simultaneously increase her profit while incentivizing higher effort from the agents than they would otherwise exert outside the contracts, thereby leading to improved state of security. Accurate pre-screening enables the insurer to take advantage of this opportunity. In the next chapter we will put this concept in a much more realistic context by following a practical underwriting framework. In particular, we will see more concretely how this positive result is rooted in the insurer's improved ability to control the agents' risks when the latter are interdependent.

3.7   TABLES OF NOTATIONS USED IN THIS CHAPTER

Table 3.1: Notations used in the two-agent model

Symbol	Definition
$e_i, e_{-i}, i = 1, 2$	One and the other agent's efforts in a two-agent system
$L^{(i)}_{e_1, e_2}$	Loss experienced by agent i given both agents' effort
x	Interdependence factor
c_1, c_2	Agent's cost of effort
γ_1, γ_2	Agent's risk attitude
S_{e_i}, W_i, σ_i^2	Agent's effort assessed by the insurer, noise in the estimate, and its variance
b_i, α_i, β_i	Agent i's policy parameters

Table 3.2: Notations used in game G^{oo}, when both agents opt out of the contract

Symbol	Definition
G^{oo}	The game between two agents, both outside the contract
$\overline{U}_i^{oo}(e_1, e_2)$	Expected utility of agent i given both agents' effort outside the contract
$B_i^{oo}(e_{-i})$	The best-response function giving i's optimal design in response to $-i$'s effort
$m_i := \arg\min_{e \geq 0} \mu(e) + c_i e$	Introduced for notational convenience
$e_i^*(m_i, m_{-i})$ or e_i^{oo}	Efforts at the NE in game G^{oo}
u_i^{oo}	Agent i's expected utility at the NE in game G^{oo}
$\tilde{e}_i$	Socially optimal effort by agent i

Table 3.3: Notations used in game G^{io}, when agent 1 opts in and agent 2 opts out of the contract

Symbol	Definition
G^{io}	The game between the insured agent 1 and uninsured agent 2
$\overline{U}_1^{io}(e_1, e_2, b_1, \alpha_1, \beta_1)$	Agent 1's expected utility in game G^{io}
$\overline{U}_2^{io}(e_1, e_2)$	Agent 2's expected utility in game G^{io}
$B_i^{io}(e_{-i})$	Agent i's best response to agent $-i$'s effort
$m_1(\alpha_1, \beta_1)$	Introduced for notational convenience in expressing the best response solution
$\overline{V}^{io}(b_1, \alpha_1, \beta_1, e_1, e_2)$	Insurer's expected utility in offering contract (b_1, α_1, β_1) to agent 1 while agent 2 opts out
u_2^{io}	Agent 2's expected utility outside insurance when the insurer offers the optimal contract to agent 1
u_1^{oi}	Agent 1's expected utility outside insurance when the insurer offers the optimal contract to agent 2

Table 3.4: Notations used in game G^{ii}, when both agents opt into contract

Symbol	Definition
G^{ii}	The game between the two agents when both opt in
$\overline{U}_1^{ii}(e_i, e_{-i}, b_i, \alpha_i, \beta_i)$	Agent i's expected utility in game G^{ii}
$B_i^{ii}(e_{-i})$	Agent i's best response to agent $-i$'s effort
$\overline{U}_i^{ii}(e_i, e_{-i}, b_i, \alpha_i, \beta_i)$	Agent i's expected utility in game G^{ii} when offered the contract (b_i, α_i, β_i)
v^{ii}	Insurer's optimal utility when both agents opt in
e_i^{ii}	Optimal effort of agent i when purchasing the optimal contract

Table 3.5: Notations used in Section 3.4

Symbol	Definition
N	Total number of agents
$(L_e^1, ..., L_e^N)$	Agents' losses, given by a multivariate Gaussian distribution
θ	Covariance value between any two agents' losses
δ	Insurer's risk attitude

3.8   APPENDIX

3.8.1   FINDING THE SOLUTION TO THE BEST-RESPONSE FUNCTION $B_i^{oo}(e_{-i})$

Below we show that the best-response function given by

$$B_i^{oo}(e_{-i}) = \arg\max_{e_i \geq 0} \ -\mu(e_i + xe_{-i}) - c_i e_i \qquad (3.25)$$

has a solution $(m_i - xe_{-i})^+$, where $m_i = \arg\min_{e \geq 0} \mu(e) + c_i e, i = 1, 2$.

First, consider the case when $m_i \geq xe_{-i}$ for the given e_{-i}. Since m_i is the minimizer of $\mu(e) + c_i e$, it follows that for any choice $e \geq 0$ we must have

$$\mu(m_i) + c_i(m_i - xe_{-i}) = \mu(m_i) + c_i m_i - c_i xe_{-i}$$
$$\leq \mu(e) + c_i e - c_i xe_{-i}. \qquad (3.26)$$

Letting $e = b + xe_{-i}$ in the above, we obtain

$$\mu(m_i) + c_i(m_i - xe_{-i}) \leq \mu(b + xe_{-i}) + c_i, \qquad (3.27)$$

which means that $e_i = (m_i - xe_{-i})^+ = m_i - xe_{-i}$ is indeed the minimizer of $\mu(e_i + xe_{-i}) + c_i e_i$, i.e., the best response.

Next, consider the case $m_i = 0 \leq xe_{-i}$. Since $e = 0$ is the minimizer to $\mu(e) + c_i e$, we must have $\mu'(0) + c_i \geq 0$, which means $\mu'(xe_{-i}) + c_i \geq 0$ by the strict convexity of $\mu()$. This means that $\mu(e_i + xe_{-i}) + c_i e_i$ is non-decreasing in e_i, and thus $e_i = (0 - xe_{-i})^+ = 0$ is the best response.

Last, consider the case $0 < m_i < xe_{-i}$. Since m_i is the minimizer of $\mu(e) + c_i e$ but not at the boundary of $e \geq 0$, the first-order optimality condition holds, meaning $\mu'(m_i) + c_i = 0$. Since $\mu'()$ is a strictly increasing function by the strict convexity of $\mu()$, and since $m_i < xe_{-i}$, we have

$$\mu'(e_i + xe_{-i}) + c_i = \mu'(e_i + xe_{-i}) - \mu'(m_i) > 0, \ \ e_i \geq 0, \qquad (3.28)$$

which means $\mu(e_i + xe_{-i}) + c_i e_i$ is increasing in e_i. This means the function is minimized at $e_i = (m_i - xe_{-i})^+ = 0$, which is also the best response.

3.8.2   FINDING A FIXED POINT

To find a fixed point to Equation (3.5), we consider three cases.

1. $e_1 = 0, e_2 \geq 0$: in this case, $e_2 = m_2$. Also, this case is valid if $m_1 - xm_2 \leq 0$, for otherwise $e_1 > 0$.

2. $e_2 = 0, e_1 \geq 0$: similar to the previous case, $e_1 = m_1$. This case is valid if $m_2 - xm_1 \leq 0$, for otherwise $e_2 > 0$.

3. $e_1 > 0, e_2 > 0$: in this case, we solve the following system of equations:

$$e_1 = m_1 - xe_2, \quad \text{and} \quad e_2 = m_2 - xe_1, \tag{3.29}$$

whose solution is given by

$$e_1 = \frac{m_1 - x \cdot m_2}{1 - x^2}, e_2 = \frac{m_2 - x \cdot m_1}{1 - x^2}. \tag{3.30}$$

Note that this case is valid if $\frac{m_1 - x \cdot m_2}{1 - x^2} > 0$ and $\frac{m_2 - x \cdot m_1}{1 - x^2} > 0$.

Putting these together, given $0 \leq x < 1$, the system of Equations (3.5) has a unique fixed point $e_i^*(m_i, m_{-i})$, i.e., the agents' effort at the unique Nash equilibrium:

$$e_i^*(m_i, m_{-i}) = \begin{cases} \frac{m_i - x \cdot m_{-i}}{1 - x^2} & \text{if } m_i \geq x \cdot m_{-i} \text{ and } \geq x \cdot m_i \\ 0 & \text{if } m_i \leq x \cdot m_{-i} \\ m_i & \text{if } m_{-i} \leq x \cdot m_i \end{cases}. \tag{3.31}$$

3.8.3   PROOF OF THEOREM 3.1 PARTS (II) AND (III)

The socially optimal efforts of agents are given by the solution to

$$\begin{aligned} (\tilde{e}_1, \tilde{e}_2) &= \arg\max_{e_1 \geq 0, e_2 \geq 0} \overline{U}_1^{oo}(e_1, e_2) + \overline{U}_2^{oo}(e_1, e_2) \\ &= \arg\max_{e_1 \geq 0, e_2 \geq 0} -\mu(e_1 + xe_2) - c_1 e_1 - \mu(e_2 + xe_1) - c_2 e_2. \end{aligned} \tag{3.32}$$

Denote by $h_i(e_{-i})$ the effort level e_i maximizing the above objective function, as a function of the other agent's effort, e_{-i}. Recall also that the optimal value of e_i as function of e_{-i} for maximizing $\overline{U}_i^{oo}$ (i.e., only agent i's utility) is given by $(m_i - xe_{-i})^+$, where $m_i = \arg\min_{e \geq 0} \mu(e) + c_i e$. We next show that $h_i(e_{-i}) \geq (m_i - xe_{-i})^+$.

We do so by contradiction. Assume that $h_i(e_{-i}) < (m_i - xe_{-i})^+$ for a given value of e_{-i}. Note that $\overline{U}_{-i}^{oo}$ is an increasing function in e_i. Also, $\overline{U}_i^{oo}$ is maximized at $e_i = (m_i - xe_{-i})^+$. As a result, with $e_i = (m_i - xe_{-i})^+$ instead of $h_i(e_{-i})$, both $\overline{U}_1^{oo}(e_i, e_{-i})$ and $\overline{U}_2^{oo}(e_i, e_{-i})$ would increase, which in turn implies that $h_i(e_{-i})$ is suboptimal, a contradiction. Therefore, $h_i(e_{-i}) \geq (m_i - xe_{-i})^+$.

Next, since $(\tilde{e}_1, \tilde{e}_2)$ solves the optimization problem (3.32), we have $h_i(\tilde{e}_{-i}) = \tilde{e}_i \geq (m_i - x\tilde{e}_{-i})^+ \geq m_i - x\tilde{e}_{-i}$. Therefore, we have

$$\tilde{e}_i \geq m_i - x\tilde{e}_{-i} \implies \tilde{e}_i + x\tilde{e}_{-i} \geq m_i \implies \tilde{e}_i + \tilde{e}_{-i} \geq m_i.$$

In other words, network security in the socially optimal solution is higher that both m_1 and m_2. In addition, we have

$$\tilde{e}_1 \geq m_1 - x\tilde{e}_2 \implies \tilde{e}_2 \geq m_2 - x\tilde{e}_1 \implies \tilde{e}_1 + \tilde{e}_2 \geq \frac{m_1 + m_2}{1 + x}.$$

That is, network security in the socially optimal solution is higher than $\frac{m_1+m_2}{1+x}$.

Recall e_1^{oo}, e_2^{oo} denote the agents' efforts at the Nash equilibrium when both are outside the contract. By (3.6), we know that

$$e_1^{oo} + e_2^{oo} = \begin{cases} m_1 & \text{if } xm_1 \geq m_2 \\ m_2 & \text{if } xm_2 \geq m_1 \\ \frac{m_1+m_2}{1+x} & \text{o.w.} \end{cases}$$

We have shown that $\tilde{e}_1 + \tilde{e}_2 \geq \max\{m_1, m_2, \frac{m_1+m_2}{1+x}\}$. Therefore, $\tilde{e}_1 + \tilde{e}_2 \geq e_1^{oo} + e_2^{oo}$. This establishes part (ii) of the theorem.

To prove part (iii) of the theorem, first we show that u_2^{io} is lower then u_2^{oo}. Recall that u_2^{io} denotes the utility of agent 2 when he is outside the contract while agent 1 is given an optimal contract; u_2^{oo} is the utility of the second agent when both agents opt out. With agent 1 inside the contract and agent 2 outside the contract, the insurer's problem is as follows:

$$\max_{\alpha_1, \beta_1, e_1, e_2} \quad -u_1^{oo} - \mu(e_1 + xe_2) - c_1 e_1 \tag{3.33}$$

$$\text{s.t.} \quad m_1(\alpha_1, \beta_1) = \arg\min_{e \geq 0} (1 - \beta_1)\mu(e) + (c_1 - \alpha_1)e$$

$$m_2 = \arg\min_{e \geq 0} \mu(e) + c_2 e$$

$$e_1 = e_1^*(m_1(\alpha_1, \beta_1), m_2)$$

$$e_2 = e_2^*(m_2, m_1(\alpha_1, \beta_1)).$$

We note that the objective function of the principal's problem is independent of α_1, β_1. As a result, we can select an optimal contract $\alpha_1 = c_1$ and $\beta_1 = 1$, in which case any non-negative effort level satisfies agent 1's IC constraint. We therefore substitute the first constraint in (3.33) with $m_1(c_1, 1) \geq 0$.

We next note that under $(\alpha_1 = c_1, \beta_1 = 1)$, any effort level e_1, and the corresponding best response of the second agent to e_1, will constitute a Nash equilibrium. We therefore re-write the principal's problem as:

$$\max_{m_1(c_1,1)\geq 0} \quad -u_1^{oo} - \mu(\max\{m_1(c_1,1), xm_2\})$$

$$-c_1(\min\{\frac{m_1(c_1,1)-xm_2}{1-x^2}, m_1(c_1,1)\})^+ \qquad (3.34)$$

$$\text{s.t.} \quad m_1(c_1,1) \geq 0$$

$$m_2 = \arg\min_{e\geq 0} \mu(e) + c_2 e.$$

Let $m_1 = \arg\min_{e\geq 0} \mu(e) + c_1 e$. We show that m_1^*, the solution to the optimization problem (3.34), is no higher than m_1.

To show $m_1^* \leq m_1$, we proceed by contradiction. Assume $m_1^* > m_1$, i.e., the first agent exerts strictly higher effort when he enters the contract (game G^{io}) than he would in the no-insurance equilibrium. We consider three cases.

1. $m_1^* > \frac{m_2}{x}$: In this case, the objective function in (3.34) is given by $-u_1^{oo} - \mu(m_1^*) - c_1 m_1^*$. The first derivative of this function is $-\mu'(m_1^*) - c_1 < 0$ since $m_1 = \arg\min_{e\geq 0} \mu(e) + c_1 e \Rightarrow \mu'(m_1) + c_1 \geq 0$ (could be positive when $m_1 = 0$) $\Rightarrow \mu'(m_1^*) + c_1 > 0$ (by the strict convexity of $\mu(.)$). Therefore, m_1^* is not optimal in this case, as its decrease improves the objective value. We thus conclude that under the assumption of this case, we should have $m_1^* \leq m_1$.

2. $\frac{m_2}{x} \geq m_1^* > x \cdot m_2$: In this case, the objective function of (3.34) is $-u_1^{oo} - \mu(m_1^*) - c_1\frac{m_1^*-xm_2}{1-x^2}$. The first derivative is given by $-\mu'(m_1^*) - \frac{c}{1-x^2}$, which is negative since $m_1 = \arg\min_{e\geq 0} \mu(e) + c_1 e \Rightarrow \mu'(m_1) + c_1 \geq 0$ (could be positive when $m_1 = 0$) $\Rightarrow -\mu'(m_1^*) - \frac{c}{1-x^2} < 0$ (by the strict convexity of $\mu(.)$). Therefore, m_1^* is not optimal in this case as decrease in m_1^* improves the objective value. Therefore, $m_1^* > m_1$ in this case as well.

3. $x \cdot m_2 \geq m_1^*$: In this case, the first agent exerts zero effort in both the no-insurance equilibrium and in the game G^{io}. This again contradicts $m_1^* > m_1$.

We therefore conclude that $m_1^* \leq m_1$, i.e., the first agent exerts less effort when only he enters the contract, than he would at the no-insurance equilibrium. This in turn leads to lower utility for the second agent, as compared to the case when both agents are outside the contract. Therefore, we have $u_2^{oo} \geq u_2^{io}$.

Similarly, we can show that $u_1^{oo} \geq u_1^{oi}$.

Last, we show that the insurer can obtain positive profit when offering the optimal contracts. Note that we have established $u_1^{oi} + u_2^{io} \leq u_1^{oo} + u_2^{oo}$. When both agents purchase the optimal contracts, the objective value is

$$\begin{aligned} v^{ii} &= \max \quad -u_1^{oi} - u_2^{io} - \mu(e_1 + xe_2) - \mu(e_2 + xe_1) - c_1 e_1 - c_2 e_2 \\ &= -u_1^{oi} - u_2^{io} + \overline{U}_1(\tilde{e}_1, \tilde{e}_2) + \overline{U}_2(\tilde{e}_1, \tilde{e}_2) \\ &\geq \overline{U}_1(\tilde{e}_1, \tilde{e}_2) + \overline{U}_2(\tilde{e}_1, \tilde{e}_2) - \overline{U}_1(e_1^o, e_2^o) - \overline{U}_2(e_1^o, e_2^o). \end{aligned}$$

This establishes the lower bound on the profit of the insurer (part (iii) of the theorem), and concludes the proof.

CHAPTER 4

A Practical Underwriting Process

In Chapter 3 we used a contract model to highlight at a conceptual level the decision making relationship between an insurer and two risk-dependent agents. In this chapter we will show what this means in practice by delving into a model that resembles much closer the reality, and analyzing it within the context of a standard underwriting framework commonly used in the insurance industry.

Since risk dependency is again our prime focus, it helps to explain now it is treated in practice—risk dependency is generally regarded as an undesirable thing to be avoided by practitioners when looking across policies in their portfolios. To understand this aversion, there are two main reasons. First, it is more likely that simultaneous loss events could happen to interdependent agents, which would threaten the insurer's capital limit or other liquidity requirements. Second, in the event that a data breach or other loss events could be attributed to a third party, such as a service provider (e.g., a cloud platform vendor), who may be insured by a *different* carrier, the insurer of the primary party may seek to recover some or all of her losses from the third party's insurer/policy, thereby reducing her own risk exposure. If, on the other hand, the primary party's insurer underwrites both the primary firm and his third party, then even if the loss to the primary could be attributed to the third party, the insurer would effectively be "suing herself" for the losses. Both these concerns seem perfectly sensible and have led to a strong desire among insurance carriers to minimize this type of risk dependency among her portfolio clients.

However, we saw from the previous chapter that contrary to this common dependency avoidance practice, there is an unrealized incentive for an insurer to underwrite dependent risks. Paradoxically, the existence of risk dependency among a network of insureds allows the insurer to jointly design polices that incentivize the insureds to (collectively) commit to higher levels of effort, which can simultaneously result in improved state of security for all as compared to a portfolio of independent insureds, and in improved profits for the insurer. It is thus of considerable interest to cyber insurance underwriters to understand how to effectively manage not only individual firm risk, but overall portfolio risk in the presence of dependent risks among policy holders, and to do so in a realistic underwriting setting.

This chapter sets out to translate the previous chapter's results into more practical terms. In particular, our next model will focus on a one-way directionality of the risk dependence, that of a set of customers relying on a common service provider (SP). This type of vendor/supplier relationship is much more common in reality than the generic mutual-dependence modeled in the previous chapter. In addition, our analysis will focus on different portfolio choices by an insurer, in terms of what collection of clients/policies to simultaneously underwrite, and quantify the impact of these choices on the resulting profit, risk reduction, as well as social welfare.

Specifically, we consider a service provider and his customers, and model three portfolio alternatives available to the insurance carrier: insure just the service provider, insure both the service provider and his customers, or insure just the service provider's customers. The strategic decision centers on how the insurer can induce the parties to reduce their risk while maximizing her own profit. We examine how these incentives can be used to reduce the direct risk to one party, as well as to reduce indirect risks to dependent firms. We also examine social welfare implications and use data from an actual cyber insurance policy, as well as one of the only sources of insurance claims data, to calibrate and substantiate our analysis.

Our results show that the insurer is able to achieve higher profit by insuring all agents (SP and his customers) provided she appropriately incentivizes the SP to improve his state of security. This is because risk reduction by the SP leads to risk reduction for his customers, thus the benefit has a multiplicative effect. This ultimately not only allows the insurer to take on the risk of all agents without hurting her profit, but also leads to higher social welfare.

We further examine whether these observations continue to hold when an insurer can recover a part of the loss suffered by an insured through a third-party liability clause when the loss can be attributed to another insured (the third party) underwritten by a different insurer. Even with this loss recovery as an alternative, we show that it is beneficial both from a security perspective and a profit perspective for an insurer to underwrite both insureds, precisely because this allows the insurer to control the risk dependency and incentivize both to commit to higher security efforts.

Overall, our results suggest a novel and improved approach to cyber insurance policy design that presents a new way of thinking about systemic risk and cyber risk dependency: to embrace and manage these risks, rather than avoid them. While we acknowledge the warranted caution against concurrent and correlated loss events, the emphasis of the present chapter is to highlight a definitive silver lining behind risk dependency, and an opportunity to actively work toward reducing overall cyber risks in an ever-escalating and interconnected threat landscape.

4.1   COMPUTING PREMIUMS USING BASE RATES

In this section we briefly describe a common approach to calculating cyber insurance premiums. The calculation begins by first selecting the *base premium* and a *base retention* (deductible) from previously defined lookup tables. The base premium is then modified through a sequence of *factors* that get multiplied onto the base quantities. While different carriers use different values and types of factors in their premium expression, there are a number of commonly used factors.

Below we provide an example of such a calculation using an actual cyber insurance policy (see [70] to view the full rate schedule), with methods commonly found throughout the insurance industry. First, the base premium and retention are determined using table lookup, where the asset size (for financial institutions) or annual revenue (for non-financial institutions) of the insured maps to assigned values, with both the rate and the retention amounts increasing in asset or revenue size. For instance, a financial institution of asset value up to $100M would be charged a base rate of $5,000 for a base retention of $25,000, while a firm of assets between $500M and $1B would be charged a base rate of $11,000 for a base retention of $100,000, all for a nominal coverage amount of $1M. On the other hand, a non-financial firm with annual revenue between $5M to $10M would be charged a base rate of $7,500 for a base retention of $25,000, and so on.

This base rate is then multiplied by a number of factors, with each factor modifying the base rate by roughly between −20% and +20% with a few exceptions, as discussed below.

- **Industry Factor:** Based on the type of business, an industry hazard is determined, with higher-risk businesses receiving a larger multiplier. For instance, agricultural and construction businesses receive the smallest hazard value (less risky) while web service providers receive the larger hazard value (more risky), as shown in Table 4.1.

- **Retention Factor:** This factor depends on the retention (deductible) that the insured selects (that differs from the base retention amount). Retention factor decreases as the selected retention increases, as shown in Table 4.2: it is 1.0 if the insured accepts the base retention; it exceeds 1.0 if the insured wants to lower this amount, and falls below 1.0 if the insured is willing to take a higher amount.

- **Increased Limit Factor:** This is a factor driven by the limit of the coverage: it is 1.0 if the insured accepts the default limit (corresponding to the base rate and base retention); it exceeds 1.0 if the insured wants to increase this limit, and falls below 1.0 if the insured asks for a lower coverage limit, as shown in Table 4.3.

- **Co-insurance Factor:** This factor is less than 1.0 if the insured accepts to pay a share of the payment made against a claim. The value of this factor depends on the amount of the share that the insured accepts to pay. Table 4.4 lists some of the co-insurance factors based on the co-insurance percentage.

- **First-Party Modifier Factors:** In the context of cybersecurity, these factors are meant to capture the insured's security posture. For instance, there could be separate factors assessing the presence of an information security policy, a laptop security policy, whether sensitive data is stored on web servers, or whether the insured has a disaster recovery plan, respectively. This information is typically collected through a questionnaire filled out as part of the insurance application process: each category consists of multiple Yes–

Table 4.1: Industry hazard table

Industry	Factor
Agriculture	0.85
Construction	0.85
Not-for-profit organizations	1.00
Technology service providers	1.2
Telecommunications	1.2

Table 4.2: Retention factor

Selected Retention ($)	Base Retention ($)			
	25,000	100,000	500,000	1000,000
25,000	**1.00**	1.16	1.34	1.47
100,000	0.87	**1.00**	1.16	1.27
500,000	0.75	0.87	**1.00**	1.10
1,000,000	0.68	0.79	0.91	**1.00**

Table 4.3: Increased limit factor

Coverage Limit ($)	Increased Limit Factor
1,000,000	1.000
2,500,000	1.865
5,000,000	2.987
10,000,000	4.786
25,000,000	8.925

Table 4.4: Co-insurance factor

Co-insurance (%)	Co-insurance Factor
0	1.000
1	0.995
5	0.980
10	0.960
20	0.920
50	0.780

No questions;[1] how many questions are answered in the affirmative then determines the value of that corresponding modifier factor.[2]

- **Third-Party Modifier Factor:** This factor depends on the third-party service provider. If the insured does not use any third-party service, this factor is 1.0. Otherwise, this factor is set based on the third-party service and the agreement between the insureds and the service provider, but is *not* a function of the security posture of the third party in the context of cyber insurance.

- **Optional Coverage Grants:** In addition to the base coverage, the policy holder may purchase coverage for additional exposures, such as privacy costs or crisis management. Each additional coverage is priced as a fraction of the base rate (premium), calculated using a number of factors including an option-specific modifying factor. For instance, the option of privacy notification expense uses a factor of 0.15, while the option of crisis management expense uses a factor of 0.02.

Other carriers use similar frameworks for calculating the final premium. We refer the interested reader to [101] for a more complete overview of current insurance practices. The multiplicative formula described above constitutes the basic model used in our analysis throughout this chapter.

Example.   We complete this section by providing an example of how the final premium is calculated using the above tables. Consider a non-financial Technology Service Provider with annual revenue $6M who intends to purchase an insurance policy with retention $100,000, coverage limit $2.5M, and zero percent co-insurance. Moreover, this firm does not use any third party services; it wishes to opt in for additional coverage for privacy notification and crisis management expenses. Based on the above tables, the following factors will be used in determining the total premium for this company:

- Base premium: $7,500; Base Retention: $25,000 (from initial table look-up, using annual revenue, for a nominal $1M in coverage).

- Industry Factor: 1.2 (Table 4.1).

- Retention Factor: 0.87 (Table 4.2).

- Limit Factor: 1.865 (Table 4.3).

- Third-Party Modifier Factor: 1.

- Co-insurance Factor: 1 (Table 4.4).

---

[1]Example questions: do you have such a policy, it is kept current and reviewed at least annually, etc.
[2]Example factor values: answering Yes to 2/1/0 of the questions results in a factor value range (0.8–0.9) / (0.95–1.05) / (1.1–1.2).

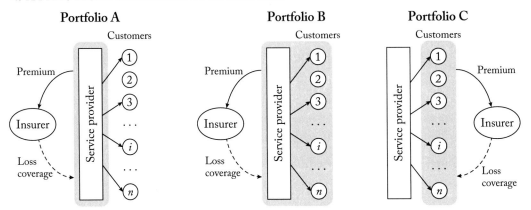

Figure 4.1: Three portfolio types: shaded areas indicate entities insured by a given underwriter.

- Privacy notification: 0.15.

- Crisis management: 0.02.

Therefore, the premium for this service provider is calculated as follows,

$$\begin{aligned} \text{Premium} &= 7500 \times 1.2 \times 0.87 \times 1.865 \times 1 \times 1 + 7500 \times (0.15 + 0.02) \\ &= 14602.95 + 1275 = \$15,877.95. \end{aligned} \tag{4.1}$$

4.2 THE INSURANCE POLICY MODEL AND ANALYSIS

We model three portfolio alternatives available to the insurance carrier, as depicted in Figure 4.1: insure only the SP and let someone else insure its customers (*Portfolio type A*), insure both the SP and its customers (*Portfolio type B*), or insure only the SP's customers and let someone else insure the SP (*Portfolio type C*).

In each case we are interested in understanding to what extent the insurer may be able to induce the parties to reduce their risk while maximizing her own utility/profit. We will examine how these policy incentives can be used to reduce the direct and indirect risks to the parties involved. To do so, over the next few subsections we develop a model that formally establishes an insurance carrier's profit as a function of the insurance policy terms as well as incentives embedded in the policy.

4.2.1 BASE PREMIUM CALCULATION

Consider an insurer and her prospective insureds (the applicants), which include a service provider (denoted by the index $i = 0$) and his n customers (denoted by $i = 1, \cdots, n$). The insurer charges a base premium b_o to the service provider and base premiums b_i to his customers $i, i = 1, 2, \cdots, n$.

Following the process outlined in the previous section, the base premium b_i depends on the total assets or revenue of the insureds. The insurer then asks the applicants to fill out a questionnaire describing their information security practices. Based on the completed questionnaire, the insurer modifies the base premiums by a (security modifier) factor f_i, $i = 0, 1, \cdots, n$. The insured pays $b_i f_i$ up front, and the insurer pays the insured $\max\{L_i - d_i, 0\}$ after a loss incident where L_i is a random variable denoting the loss amount of agent i and d_i is his elected retention/deductible. For the analysis that follows we shall ignore all the other factors unrelated to cybersecurity, as their inclusion (as additional multipliers) does not affect our model or our conclusion.

Note that insured i's premium $b_i f_i$, $i = 1, \cdots, n$, is purely a function of his own security posture. While the information security questionnaire used to generate modifier factor f_i may include questions on whether i has a third-party supplier, or whether he has proper procedures/policies in place that govern the handling of a third party, it does not directly assess the security posture of these third parties themselves. We refer an interested reader to the Chubb CyberSecurity policy given in the Appendix of [70].

4.2.2   THE SECURITY INCENTIVE MODIFIER

We now introduce an incentive factor, f_o', for the SP, and subsequently examine its impact on the SP as well as his n customers. Specifically, suppose the insurer is willing to offer the SP a discounted premium in exchange for improved security posture as follows.

- The SP has an initially assessed premium $b_o f_o$, with a security modifier factor f_o.

- The SP agrees to invest more in security such that he could now be assessed at $\tilde{f}_o = f_o - f_o'$, for some $f_o' \in [0, f_o]$, i.e., a reduction in the modifier factor.

- In return, the insurer agrees to revise the premium to $b_o \tilde{f}_o$, reflecting a discount given the SP's improved security posture. In other words, $b_o f_o'$ is the discount the SP receives.

Note that here for simplicity of presentation, we have assumed that the insurer is able to assess, and willing to match exactly in discount the amount corresponding to the reduced risk. That is, this SP now enjoys a revised premium equal to that which he would have received had he started at a security level measured at $\tilde{f}_o$ without the incentive. In practice, the two need not be equal, i.e., the SP may require more or less in premium discount incentive to reach $\tilde{f}_o$. While this does not affect our qualitative conclusions, it does raise the interesting question as to whether in practice the incentive offered is sufficient for the SP to attain the corresponding risk reduction. In other words, could the SP take the discount $b_o f_o'$ and use it toward hiring additional personnel or purchasing products to achieve this goal? We will give such an example in Section 4.3.

Our subsequent analysis focuses on whether a desirable operating point exists for the insurer to offer an incentive to the SP ($f_o' > 0$). Obviously, when there is no incentive ($\tilde{f}_o = f_o$), the problem reverts to the original premium calculation. Note that we are singularly focused on incentivizing the SP—while an insurer can obviously also provide the same type of incentive to the SP's customers, we will not include this in our model as a customer's risk does not spill over to others in this system as illustrated in Figure 4.1.

4.2.3   MAPPING SECURITY INCENTIVE TO PROBABILITY OF LOSS

The security modifier factor f_i is tied to some underlying assumption of the probability of a cyber incident. This modifier can increase or decrease the base premium; the larger it is, the more likely is a loss event as estimated by the insurer. To the best of our understanding, by examining the rate schedules of many actual cyber insurance policies, this factor himself is not directly tied to the *magnitude* of a loss. Rather, we believe the expected loss amount is factored into the base premium which is itself tied to the sector/industry and the size of the insured. The use of such a factor in the current underwriting practice would suggest that policies are risk priced in additional to being market priced (reflected in the base premium and retention). This aspect, however, does not affect our analysis since we only consider a single insurer.

To be concrete, let $P_o(\tilde{f}_o)$ denote the probability of a breach to the SP, which is decreasing in the security incentive factor f_o' and increasing in the overall factor $\tilde{f}_o$. Similarly, we denote by $P_i(f_i)$, $i = 1, \cdots, n$, the probability of a loss incident of customer i *unrelated* to the SP. Both $P_o()$ and $P_i()$ are assumed to be increasing and differentiable. We will assume that if a breach happens to the SP, a business interruption or similar loss event occurs to his customer with probability t, also referred to as the level/degree of dependency. Further, we will assume that a business interruption induced by SP and the loss incident unrelated to the SP are independent events.

Putting these together, the probability of a loss event occurring to customer i is given by:

$$P_{li}(\tilde{f}_o, f_i) = P_i(f_i) + t P_o(\tilde{f}_o)(1 - P_i(f_i)), \quad i = 1, \cdots, n, \tag{4.2}$$

where the loss includes that due to the customer himself, due to business interruption brought on by the SP's breach, or both at the same time. This will also be written as $P_{li}(f_o - f_o', f_i)$ to highlight the fact that f_o' is the controllable parameter.

4.2.4   THE INSURER'S UTILITY/PROFIT FUNCTION

Next, we derive expressions for the insurer's utility under two portfolio options: when she insures just the SP (Portfolio A), and when she insures both the SP and his customers (Portfolio B).

The insurer's utility (V_o) and expected utility ($\overline{V}_o$) from underwriting *only* the SP are defined as follows, both shown as functions of f_o' given that our focus is on this element under the insurer's control,

$$V_o(f_o') = b_o(f_o - f_o') - I_o(L_o - d_o)^+; \tag{4.3}$$

$$\overline{V}_o(f_o') = E[V_o(f_o')] = b_o(f_o - f_o') - l_o P_o(f_o - f_o'), \tag{4.4}$$

where $(x)^+ = \max\{x, 0\}$, and $l_o = E[(L_o - d_o)^+]$. For convenience we have used I_o to denote a Bernoulli random variable with parameter $P_o(f_o - f_o')$.

We will assume the customers' security modifier factors $f_i, i = 1, \cdots, n$, are uniformly distributed over some range $[f_{\min}, f_{\max}]$. The insurer's utility from customer i is then given by the following, again expressed as a function of the controllable f_o':

$$V_i(f_o') = b_i f_i - I_i (L_i - d_i)^+; \qquad (4.5)$$

$$\overline{V}_i(f_o') = b_i \frac{f_{\min} + f_{\max}}{2} - \overline{P}_{li}(f_o - f_o', f_i)l_i, \qquad (4.6)$$

where $l_i = E[(L_i - d_i^+]$ and $\overline{P}_{li}$ is the expected value of P_{li} with respect to the distribution of f_i. Again, I_i denotes a Bernoulli random variable with parameter $P_{li}(f_o - f_o', f_i)$.

If the insurer chooses to underwrite *both* the SP and his n customers then her expected total utility is given by:

$$\overline{V}_{total}(f_o') = \overline{V}_o(f_o') + \sum_{i=1}^{n} \overline{V}_i(f_o'); \qquad (4.7)$$

$$\overline{V}_{\max} = \max_{f_o'} \overline{V}_{total}(f_o'). \qquad (4.8)$$

4.2.5 ANALYSIS OF THE OPTIMAL INCENTIVES AND CARRIER UTILITY/PROFIT

Now that we have established expressions for the carrier's utility as a function of security incentives, we next seek to answer two questions: first, what security incentives should the carrier provide the service provider, and secondly, which portfolio strategy yields higher utility?

Recall $P_o(\tilde{f}_o)$ is an increasing function of $\tilde{f}_o$ by definition, implying that $P_o(f_o - f_o')$ is a decreasing function of the incentive f_o'. We assume this to be a strictly convex function of f_o', reflecting a decreasing marginal return on effort. Note that it is widely accepted to model loss probability as a function of the security investment, see, e.g., [63, 82, 88, 96]. Our model here is consistent with this literature since we have assumed that the incentive factor f_o' is proportional to security effort or investment, while allowing us to highlight and express this function in terms of the carrier's controllable in this underwriting framework.

Our first result compares the optimal incentive that an insurance carrier would offer the SP when insuring just the SP (Portfolio A) vs. insuring both the SP and its customers (Portfolio B). That is, we compare the optimal incentive factor f_o^* that maximizes $\overline{V}_o()$ with the optimal incentive factor f_o^{**} that maximizes $\overline{V}_{total}()$.

Theorem 4.1 *Under the assumption that $P_o(f_o - f_o')$ is decreasing and strictly convex in f_o', we have $f_o^* \leq f_o^{**}$, where $f_o^* = \arg\max_{f_o'} \overline{V}_o(f_o')$ and $f_o^{**} = \arg\max_{f_o'} \overline{V}_{total}(f_o')$. In other words, the underwriter offers a higher incentive to the SP when insuring all parties, compared to the incentive offered to the SP as the only insured.*

Proof. The insurer's utility in underwriting the SP and his customers is given by:

$$\overline{V}_{total}(f_o') \;=\; \overline{V}_o(f_o') + \sum_{i=1}^{n} \overline{V}_i(f_o')$$

$$=\; b_o(f_o - f_o') - l_o P_o(f_o - f_o') + \sum_{i=1}^{n} b_i \frac{f_{min} + f_{max}}{2}$$

$$-l_i t P_o(f_o - f_o')(1 - \overline{P}_i(f_i)) - l_i \overline{P}_i(f_i), \qquad (4.9)$$

where $\overline{P}_i(f_i) = E[P_i(f_i)]$. Using the first order optimality condition, we have

$$\frac{\partial \overline{V}_{total}(f_o')}{\partial f_o'} = 0 \qquad (4.10)$$

$$\Rightarrow \quad f_o^{**} = \left(f_o - (P_o')^{-1}\left(\frac{b_o}{\left[l_o + \sum_{i=1}^{n} l_i t \left(1 - \overline{P}_i(f_i)\right) \right]} \right) \right)^{+}. \qquad (4.11)$$

Similarly, we can find the optimal value f_o^* that maximizes $\overline{V}_o$:

$$\frac{\partial \overline{V}_o}{\partial f_o'} = -b_o + l_o P_o'(f_o - f_o') = 0$$

$$\Rightarrow \quad f_o^* = \left(f_o - (P_o')^{-1}\left(\frac{b_o}{l_o}\right) \right)^{+}. \qquad (4.12)$$

Because $P_i'()$ is an increasing function and $\frac{b_o}{l_o} > \frac{b_o}{(l_o + \sum_{i=1}^{n} l_i t(1 - \overline{P}_i(f_i)))}$, we conclude that $f_o^* \leq f_o^{**}$. $\qquad \square$

Theorem 4.1 suggests that if the insurer underwrites both the SP and his customers (Portfolio B), she benefits from a better state of security (induced by higher incentive to the SP) as compared to if she only underwrites the SP (Portfolio A). Intuitively, as the SP's risk directly impacts that of his customers, when insuring both, it is in the insurer's interest to control/reduce the SP's risk so that the overall, systemic risk she is exposed to is reduced. This obviously means better overall security posture for all parties. The question is whether the insurer will voluntarily choose Portfolio B over A. The next result answers this.

Corollary 4.2 *If parameter values b_i and l_i are such that $\overline{V}_i(f_o^*) > 0$, then we have the following:*

$$\overline{V}_{total}(f_o^{**}) \underbrace{\geq}_{\text{optimality of } f_o^{**}} \overline{V}_{total}(f_o^*) \underbrace{\geq}_{\text{positivity of } \overline{V}_i(f_o^*)} \overline{V}_o(f_o^*). \qquad (4.13)$$

Similarly,

$$\overline{V}_{total}(f_o^{**}) \underbrace{\geq}_{\text{optimality of } f_o^{**}} \overline{V}_{total}(f_o^*) \underbrace{\geq}_{\text{positivity of } \overline{V}_i(f_o^*)} \overline{V}_i(f_o^*) \geq \overline{V}_i(0), \qquad (4.14)$$

where the last inequality results from the fact that the risk sustained by customer i is lower when the SP is incentivized at any level $f_o^* > 0$, and all the other inequalities are as explained.

The above result suggests that at the right level of incentive for the SP, the insurer enjoys greater utility by insuring both the SP and his customers (Portfolio B), relative to insuring just the SP (Portfolio A), or any subset of his customers.

The condition $\overline{V}_i(f_o^*) > 0$ needed for the corollary means there is positive expected utility from any single policy when the SP is incentivized at the level f_o^*. This need not be true if b_i is too small and l_i too large, in which case a rational insurer would not underwrite the policy.

Note that Theorem 4.1 remains valid even when the assessment is noisy. To see this, let us assume that the SP is assessed at $\tilde{f}_o = f_o - f_o'$, but the true value is $\tilde{f}_o + W$, where W is a zero-mean random variable. Then we have:

$$\overline{V}_o(f_o') = b_o \cdot (f_o - f_o') - E[P_o(f_o - f_o' + W)] \cdot l_o. \tag{4.15}$$

Thus, as long as the function $E[P_o(f_o - f_o' + W)]$ is increasing and convex, the result of Theorem 4.1 is valid. We next show that this is indeed an increasing and convex function. For simplicity of exposition, we will denote this function as $\Pi_o(f_o - f_o') = E[P_o(f_o - f_o' + W)]$, and denote the pdf of W by $g(.)$.

$$\Pi_o(x) = \int P_o(x + s)g(s)ds \quad \rightarrow \quad \Pi_o'(x) = \int \underbrace{P_o'(x + s)}_{P_o(.) \; increasing}g(s)ds \quad \geq \quad 0. \tag{4.16}$$

$$\Pi_o(\lambda x + (1 - \lambda)y) = \int \underbrace{P_o(\lambda x + (1 - \lambda)y + s)}_{convexity \; of \; P_o(.)}g(s)ds \quad \leq$$

$$\int \lambda P_o(x + s)g(s)ds + \int (1 - \lambda)P_o(y + s)g(s)ds = \lambda \Pi_o(x) + (1 - \lambda)\Pi_o(y). \tag{4.17}$$

4.2.6   PORTFOLIO C AND THIRD-PARTY LIABILITY CLAUSES

Next consider Portfolio C. In this case the SP is insured by another carrier, referred to as the *third-party insurer*. The insurer of the SP's customers (underwriter of Portfolio C) is accordingly referred to as the primary-party insurer.

Third-party liability refers to the ability of an injured party to seek redress for losses from an injurer, and is a coverage category commonly found in insurance policies. In our context, this means that if a firm suffers loss due to business interruption brought on by a breach at his SP, the firm's insurance carrier can, on the firm's behalf, seek redress from the SP's insurer (the third-party insurer). This is illustrated in Figure 4.2 in the case of Portfolio C, where the thick dashed curve shows that in the event of losses suffered by the customers of the SP due to SP's

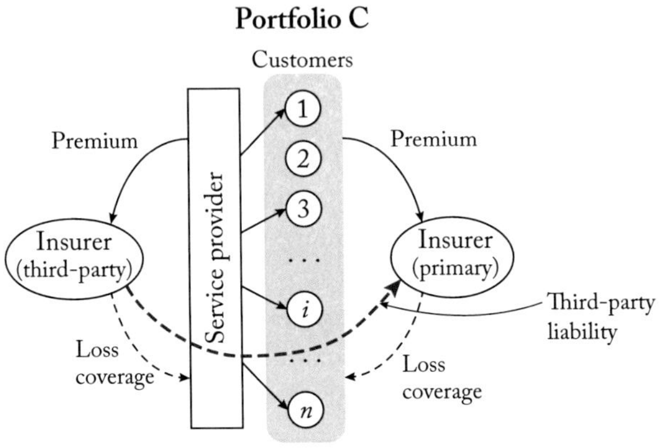

Portfolio C

Figure 4.2: Third-party liability: the SP's insurer may be liable for some or all of the damages suffered by the SP's customers.

failure, the third-party insurer may have to pay damages to the primary insurer, who then uses it to cover the customers' losses.[3]

In one of the few datasets that report actual cyber insurance claims data, NetDiligence [46] shows that 13% of all data breaches and cyber incidents can be attributed to a third party. We will use a parameter q to represent the probability that a loss can be attributed to an SP.

Denote by U the insurer's utility when she underwrites only the SP's customers (Portfolio C). We have:

$$
\begin{aligned}
U_i(f_o') &= b_i f_i - J_i(L_i - d_i)^+, \\
\overline{U}_i(f_o') &= E[U_i(f_o')] \\
&= b_i \frac{f_{\min} + f_{\max}}{2} - (\overline{P}_i(f_i) + (1-q)tP_o(f_o - f_o')(1 - \overline{P}_i(f_i)))l_i, \quad (4.18)
\end{aligned}
$$

where J_i is a Bernoulli random variable with parameter $P_i(f_i) + (1-q)tP_o(f_o - f_o')(1 - P_i(f_i))$; this is the probability that a loss incident happens to customer i and *cannot* be attributed to the SP. The third-party insurer's utility is given by:

$$
U_o(f_o') = b_o(f_o - f_o') - I_o(L_o - d_o)^+ - \sum_{i=1}^{n} K_i(L_i - d_i)^+; \quad (4.19)
$$

$$
\begin{aligned}
\overline{U}_o(f_o') &= E[U_o(f_o')] \\
&= b_o(f_o - f_o') - P_o(f_o - f_o')l_o - \sum_{i=1}^{n} qtP_o(f_o - f_o')(1 - \overline{P}_i(f_i))l_i, \quad (4.20)
\end{aligned}
$$

---

[3]In the case of Portfolio B, the same carrier underwrites both the firm and the SP. Thus, even though this compensation may be counted against the SP's policy, it makes no difference to the insurer in terms of her utility.

where K_i is a Bernoulli random variable with parameter $qtP_o(f_o - f_o')(1 - P_i(f_i))$; this is the probability that a loss incident happens to customer i and *can* be attributed to the third party successfully. Here we have assumed that whenever losses can be attributed to the SP, the customer's insurer (the primary) is fully reimbursed. However, our result in Theorem 4.3 below remains valid for partial or fractional compensation as well.

It is worth pointing out a key and interesting observation on Portfolio C, which is that the profit is split between the primary's insurer (who is underwriting the customers) and the third-party insurer (who is underwriting the SP), but the total remains unchanged from Portfolio B when everything is underwritten by the same insurer, provided the same incentive f_o' is applied. In other words, we have

$$\overline{V}_{total}(f_o') = \overline{V}_o(f_o') + \sum_{i=1}^{n} \overline{V}_i(f_o') = \overline{U}_o(f_o') + \sum_{i=1}^{n} \overline{U}_i(f_o'). \tag{4.21}$$

To see why the second equality holds, we first note that by using Equation (4.4) we have

$$\overline{U}_o(f_o') = \overline{V}_o(f_o') - \sum_{i=1}^{n} qtP_o(f_o - f_o')(1 - \overline{P}_i(f_i))l_i. \tag{4.22}$$

At the same time, using Equation (4.6)

$$
\begin{aligned}
\overline{U}_i(f_o') &= b_i \frac{f_{min} + f_{max}}{2} - (\overline{P}_i(f_i) + (1 - q)tP_o(f_o - f_o')(1 - \overline{P}_i(f_i)))l_i \\
&= b_i \frac{f_{min} + f_{max}}{2} - (\overline{P}_i(f_i) + tP_o(f_o - f_o')(1 - \overline{P}_i(f_i)))l_i \\
&\quad + tqP_o(f_o - f_o')(1 - \overline{P}_i(f_i))l_i \\
&= \overline{V}_i(f_o') + tqP_o(f_o - f_o')(1 - \overline{P}_i(f_i))l_i. \tag{4.23}
\end{aligned}
$$

The last terms in (4.23), $i = 1, \cdots, n$, and (4.22) cancel out when added, yielding the equality in (4.21).

We are now ready to compare the insurer's utility from underwriting only the SP's customers (with the possibility of recovering losses from the SP's insurer) (Portfolio C) with its utility in underwriting both the SP and its customers (Portfolio B).

We denote the insurer's utility from underwriting Portfolio C, only the SP's customers, as $\overline{U}_{max} = \sum_{i=1}^{n} \overline{U}_i(f_o^\star)$, where $f_o^\star = \arg\max_{f_o'} \overline{U}_o(f_o')$, and recall the insurer's maximum utility from underwriting both the SP and its customers is $\overline{V}_{max}$ from Equation (4.8), where the maximum is attained at f_o^{**}.

Theorem 4.3 *At the right level of incentive for the SP, the insurer enjoys greater utility by insuring both the SP and its customers (Portfolio B), rather than just the SP's customers (Portfolio C). That is, $\overline{V}_{max} \geq \overline{U}_{max}$, where $\overline{V}_{max} = \overline{V}_o(f_o^{**}) + \sum_{i=1}^{n} \overline{V}_i(f_o^{**})$, and $\overline{U}_{max} = \sum_{i=1}^{n} \overline{U}_i(f_o^\star)$. Moreover, given that $P_o(f_o - f_o')$ is decreasing and convex in f_o', we have $f_o^\star \leq f_o^{**}$, i.e., the state of security improves for both the SP and its customers when the insurer underwrites both.*

The proof can be found in Appendix 4.7. The first part of this result is rather trivial: if the primary insurer is compensated by the third-party insurer, it must therefore be profitable to underwrite the SP (otherwise the SP would not be able to obtain a policy in the first place). Thus, the primary insurer can only gain by insuring the SP as well.

The second part of the result is more interesting and less straightforward. The intuition is that when the insurer underwrites both the SP and his customers (Portfolio B), it is in her best interest to provide stronger incentive to the SP in an attempt to reap the multiplicative effect (i.e., the positive externality) of risk reduction of the SP on his customers. In short, then, by embracing risk dependency, the insurer not only gains in profit but also contributes to overall risk reduction.

As in the case of Theorem 4.1, the result of Theorem 4.3 remains valid even when the SP's assessment is noisy, by following the same argument.

4.2.7   SUMMARY OF OUR FINDINGS

The findings suggested by the analysis shown in this section are summarized as follows.

- Given the choice between insuring just the SP (Portfolio A), or the SP and all its customers (Portfolio B), an insurance carrier should choose Portfolio B. The reason is that the insurer can incentivize the SP to improve its security posture in exchange for discounted premium. While this reduces the insurer's revenue from the SP, it improves the security posture of the SP and his customers, leading to fewer claims from business interruptions. Collectively this leads to lower overall risk and higher utility/profit for the insurer.

- Given the choice between insuring both the SP and its customers (Portfolio B), or just the SP's customers (Portfolio C) and attributing losses to the SP, an insurance carrier should choose Portfolio B. This is because with Portfolio C the insurer is unable to effectively induce the SP to improve its security posture, which negatively affects all of the SP's customers.

- If an insurer chooses to underwrite only the SP's customers (Portfolio C), it should incorporate the risk condition of the SP into his customers' premiums. By contrast, current practice often ignores the security posture of the SP (or any third parties) when pricing the customer's policy.

Next, we use data from an actual cyber insurance policy, as well as one of the few sources of insurance claims data, to calibrate and substantiate our analysis through numerical examples.

4.3   NUMERICAL EXAMPLES

In this section we examine closely a number of numerical examples that put the preceding analytical results into perspective. To do so, we will need to substantiate two elements of our

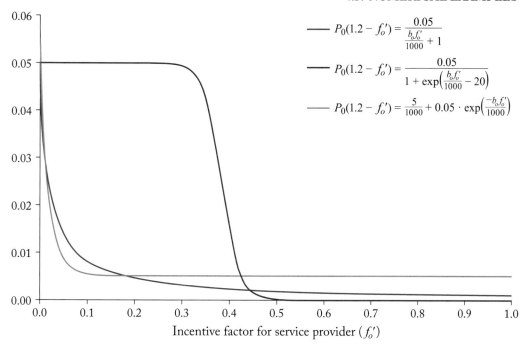

Figure 4.3: $P_o(1.2 - f_o')$, the probability of a loss event to the SP.

model: the relationship between the security modifying factor, i.e., the function $P(f)$, and the loss distribution governing L. We will also use base premium and retention values presented in Section 4.1.

4.3.1 EXAMPLES OF THE LOSS PROBABILITY FUNCTION

We present three examples of $P_o(f_o - f_o')$ as a function of f_o' while fixing $f_o = 1.2$ and $b_o = 52,000$; these are illustrated in Figure 4.3 and used in this section to perform numerical analysis.

$$P_o(f_o - f_o') = \frac{0.05}{\frac{b_o(1.2 - (f_o - f_o'))}{1000} + 1} \tag{4.24}$$

$$P_o(f_o - f_o') = \frac{0.05}{(1 + \exp(\frac{b_o \cdot (1.2 - (f_o - f_o'))}{1000} - 20))} \tag{4.25}$$

$$P_o(f_o - f_o') = \frac{5}{1000} + 0.05 \cdot \exp\left(-\frac{b_o \cdot (1.2 - (f_o - f_o'))}{1000}\right). \tag{4.26}$$

The choice of these functions are somewhat arbitrary: the main intent is to capture a few families of decreasing functions with subtle yet significant differences as explained below, while noting that our conclusion and results hold more generally. Specifically:

- The loss given in Equation (4.24) (the blue curve) is simply a decreasing, convex function which indicates that initial effort in risk reduction results in large marginal benefits, and that the loss probability will continue to decrease, albeit at a diminishing rate. This would apply to a typical firm whose initial investment (say in firewall) is very effective, after which more expensive products (e.g., intrusion detection) continue to reduce risk but at a decreasing rate.

- The loss in Equation (4.25) (the red curve) suggests the initial effort has to be significant enough (exceeding a threshold) to have any appreciable effect on loss reduction. Equivalently, this may be viewed as modeling a type of firms that only respond to incentives when they are substantial or when they reach a tipping point. Beyond this, the curve similarly exhibits diminishing returns. Note that this loss function is not convex but the result of Theorem 4.1 holds in this case as well [70].

- Finally, the loss in Equation (4.26) (the yellow curve) illustrates a scenario where the reduction in loss initially behaves similarly to the first case, but hits a ceiling beyond which no amount of effort can further reduce. This captures the situation where external factors beyond the insured's control is at significant play, contributing to a non-zero "floor" in the probability of a loss event. This could apply to the case where there is persistent susceptibility to social engineering that no amount of investment or training can completely remove; or, where the firm is simply not able to address all security challenges.

It should be noted that the above examples serve to illustrate the different ways loss probabilities may change as incentives/security investments increase. The actual values used may or may not accurately reflect reality. For instance, in reality the scale of the loss probability could be orders of magnitude larger (0.1 instead of 0.01) or smaller (0.001 instead of 0.1). Unfortunately, there is no publicly available data that would allow us to calibrate. As already mentioned, it is unclear how these factor values were derived by an underwriter in the first place.

4.3.2 EXAMPLES OF THE LOSS DISTRIBUTION

We will use data reported in the cyber insurance claims study by NetDiligence [46] to obtain breach loss distributions, summarized in Table 4.5. The "Mid Revenue" range contains somewhat unexpected small median and mean values. This appears to be an anomaly: since the sample sizes (number of cases) are small, an oversized or undersized breach can significantly throw off the average.

Table 4.5: Cost of data breach between 2016–2017, organized based on the breached firm's revenue

	Cases	Median ($)	Mean ($)
Nano revenue (< $50M)	52	49,000	215,297
Micro revenue ($50M–$300M)	31	88,154	487,411
Small revenue ($300M–$2B)	15	118,671	599,907
Mid revenue ($2B–$10B)	9	91,457	173,851
Large revenue ($10B–$100B)	8	3,326,313	5,965,571

4.3.3   EXAMPLE 1: AN SP AND A CUSTOMER WITH LARGE REVENUE

In this example, we consider a SP and a single customer, both of large revenue (e.g., a major web hosting provider and a large corporate customer). Using the rate schedule provided in Section 4.1, we will set the base premium and base retention for the SP and his customer to be $b_o = b_1 = \$52,000$ and $d_o = d_1 = \$250,000$, respectively. We consider the following loss function for the customer: $P_1(f_1) = \frac{0.05}{\frac{b_1 \cdot (1.2 - f_1)}{1000} + 1}$. Moreover, factor f_1 is uniformly distributed over $[0.6, 1.2]$ with $E[f_1] = 0.9$, depending on the outcome of his information security questionnaire.

Using the NetDiligence data, we will assume that both L_o and L_1 are log-normally distributed with a mean of \$5,965,571 and median \$3,326,313. Moreover, as mentioned earlier, NetDiligence reports that 13% of data breaches can be attributed to a third party; we will accordingly set $q = 0.13$. We will assume that the SP was assessed with $f_o = 1.2$.

We will first consider $P_o(f_o - f'_o) = \frac{0.05}{\frac{52000 \cdot (1.2 - (f_o - f'_o))}{1000} + 1}$, with results shown in Figure 4.4.

Figure 4.4a plots the optimal incentive factor under different portfolios as a function of the dependency factor t. Recall that:

- Under Portfolio A: $f_o^* = \arg\max_{f'_o} \overline{V}_o(f'_o)$,

- Under Portfolio B: $f_o^{**} = \arg\max_{f'_o} \overline{V}_{total}(f'_o)$,

- Under Portfolio C: $f_o^\star = \arg\max_{f'_o} \overline{U}_o(f'_o)$.

Figure 4.4a shows that, if the insurer underwrites only the SP (Portfolio A, blue line), t does not factor into the policy decision. On the other hand, if the insurer underwrites both, then offering incentive to the SP is now in its interest, and the incentives increases as t increases (Portfolio B, red line). Finally, if an insurer underwrites only the SP and pays the third-party compensation for his customer's loss (yellow line), the incentive factor is also increasing as a function of t but it increases slower than f_o^{**}.

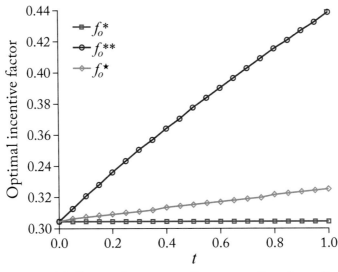

(a) Optimal incentive factor as a function of t

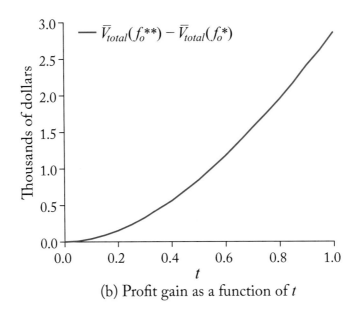

(b) Profit gain as a function of t

Figure 4.4: Optimal incentive factor and profit gain under loss model (4.24).

Figure 4.4b shows how much can be gained by taking risk dependency into account, and the higher the dependency the more the insurer gains by jointly designing contracts for both the SP and its customer.

The cases $P_o(f_o - f'_o) = \frac{0.05}{(1+\exp(\frac{b_o(1.2-(f_o-f'_o))}{1000} - 20))}$ and $P_o(f_o - f'_o) = \frac{5}{1000}$ $+ 0.05 \exp(-\frac{b_o(1.2-(f_o-f'_o))}{1000})$ are shown in Figures 4.5 and 4.6, respectively. We see the similar results as in Figure 4.4.

4.3.4 EXAMPLE 2: AN SP AND MULTIPLE CUSTOMERS WITH SMALLER REVENUE

In this example, we consider an SP and n customers with relatively small revenue and study the impact of n on the optimal policy and the insurer's utility.

Again, using the rate schedule provided in 4.1, we will set the base rate and retention for the customers at $b_i = \$5,000$, $d_i = \$25,000$, $i = 1, \cdots, n$. The factors f_i, $i = 1, \cdots, n$, are drawn uniformly from [0.6, 1.2]. Using Table 4.5, the loss random variable L_i, $i = 1, \cdots, n$, has a mean and median of \$599,907 and \$118,671, respectively. Similar as in the previous example, the mean and median of loss L_o are set at \$5,965,571 and \$3,326,313, respectively. We again assume that L_i follows a log-normal distribution. In addition, we set $f_o = 1.2$, $t = 0.5$, and $q = 0.13$. Compared to the previous example, in this example we shall also examine the effect n, the number of customers, on the optimal policy. Moreover, we consider the following loss function for customer i: $P_i(f_i) = \frac{0.05}{\frac{5000(1.2-f_i)}{1000}+1}$. The results are shown in Figure 4.7.

Figure 4.7a illustrates the optimal incentive factors f_o^*, f_o^{**}, $f_o^\star$ as a function of n. This plot shows that as the number of customers increases, the SP's insurer would incentivize the SP more under both portfolios B and C. The reasons behind this is obvious: as the risk spillover impacts more customers, the more the SP can reduce his risk, the greater the benefit to the SP's insurer. Specifically, given that a breach occurred to the SP, the probability of any downstream business interruption is given by $1 - (1 - t)^n$, which is an increasing function of n. Thus, it is in the insurer's interest to reduce the likelihood of loss on the part of the SP. As a result, both f_o^{**} and $f_o^\star$ are increasing in n, while f_o^* is independent of n as it maximizes only $\overline{V}_o$ without considering dependency.

Figure 4.7b shows that the insurer does not gain by underwriting the customers and attributing all or a part of the loss to the SP as compared to the profit by underwriting all of them. We see in some cases the third party's insurer has negative expected profit, in which case a policy is not viable.

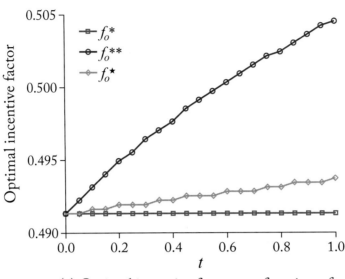

(a) Optimal incentive factor as a function of t

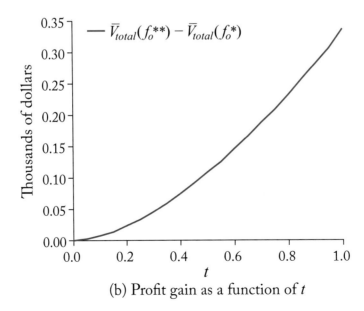

(b) Profit gain as a function of t

Figure 4.5: Optimal incentive factor and profit gain under loss model (4.25).

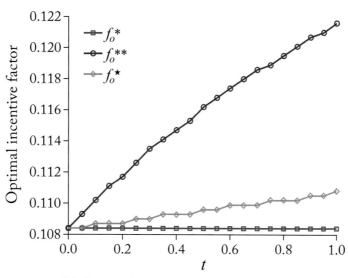

(a) Optimal incentive factor as a function of t

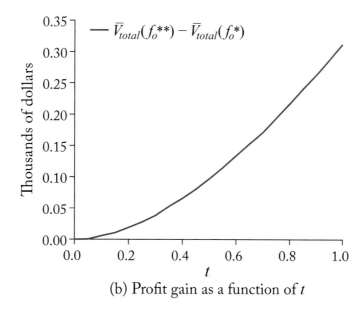

(b) Profit gain as a function of t

Figure 4.6: Optimal incentive factor and profit gain under loss model (4.26).

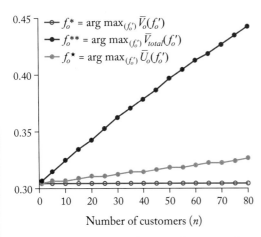

(a) Optimal incentive factor as function of n. Optimal incentive factor is increasing in n.

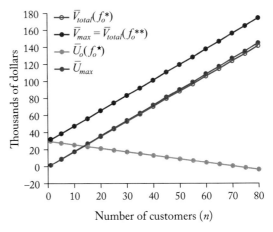

(b) Insurer's profit as function of n. The insurer does not gain by underwriting the SP's customers and attributing the loss to the SP.

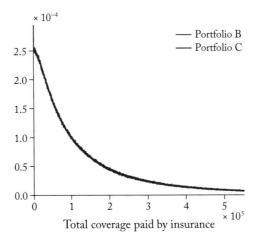

(c) Probability Distribution Function (pdf) of the amount paid out by the insurer in different scenarios.

Figure 4.7: Optimal incentive factor, insurer's profit, and pdf of the coverage paid by the insurer under loss model (4.24).

Figures 4.8 and 4.9 shows similar results for the other two loss functions

$$P_o(f_o - f_o') = \frac{0.05}{(1 + \exp(\frac{b_o(1.2 - (f_o - f_o'))}{1000} - 20))} \quad \text{and}$$

$$P_o(f_o - f_o') = \frac{5}{1000} + 0.05 \cdot \exp(-\frac{b_o \cdot (1.2 - (f_o - f_o'))}{1000}).$$

We now comment on Figures 4.7c, 4.8c, and 4.9c, which illustrate the insurer's payout distribution when the SP has $n = 10$ customers. All three show that portfolios B and C are faced with the same payout distributions regardless of the loss models used. This is in contrast to the earlier comparison when there is only a single customer. This is because more customers lead the insurer to increase her incentive for the SP in order to lower his risk and his customers' risk; this is absent under portfolio C. As a result of this, the two portfolios actually experience the same amount of risk in payout. In this sense, portfolio B is uniformly better than C.

4.4 DISCUSSIONS

We further discuss three aspects of the model studied in this chapter.

4.4.1 IS THE PREMIUM DISCOUNT SUFFICIENT?

Consider a non-financial technology service provider firm with annual revenue between $5M and $10M. In this case, the base premium $b_o = \$7,500$. We will assume the firm is assessed with $f_o = 1.2$. Now assume that the insurer sets the incentive factor f_o' to be 0.35. Therefore, the firm pays $b_o \cdot (f_o - f_o') = \$6375$ as the premium, after receiving $b_o \cdot f_o' = \$2625$ in discount. Using salary surveys such as [7], consider an IT security personnel with a bachelor's degree, 5 years of experience, and an annual salary $Wage = \$85K$ for $N = 50$ working weeks. The premium discount the firm receives can be translated into a fraction of this person's wages:

$$\frac{b_o \cdot f_o'}{Wage} \times N = \frac{\$2625}{\$85000} \times 50 = 1.5 \text{ weeks.} \tag{4.27}$$

Therefore, the incentive provided by the underwriter is just enough to hire an experienced person for 10 days. It is debatable whether this amount of investment in security is adequate to reduce the firm's cyber risk (by 10^{-9} according to Model (4.25), or by 0.05 according to Model (4.26), by setting $b_o = \$7,500$ in each, respectively). A potential mismatch between what this analysis suggests and reality may be attributed to two factors. First, as already mentioned, the loss values shown in Figure 4.3 could be orders of magnitude different from reality; in other words, if the risk reduction is from a breach probability of 0.1–0.07%, then perhaps 10 days' worth of work (say in deploying software patches) is sufficient. Secondly, it may also be argued that the current levels of base premium is inconsistent with the underlying cyber risk (and what it takes to reduce the risk) to begin with.

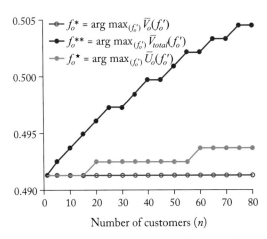

(a) Optimal incentive factor as function of n. Optimal incentive factor is increasing in n.

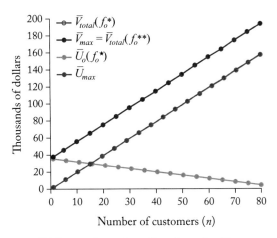

(b) Insurer's profit as function of n. The insurer does not gain by underwriting the SP's customers and attributing the loss to the SP.

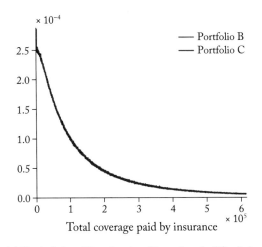

(c) Probability Distribution Function (pdf) of the amount paid out by the insurer in different scenarios.

Figure 4.8: Optimal incentive factor, insurer's profit, and pdf of the coverage paid by insurer under loss model (4.25).

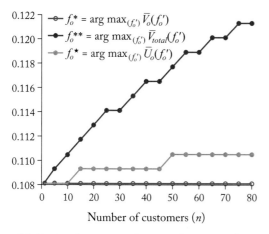

(a) Optimal incentive factor as function of n. Optimal incentive factor is increasing in n.

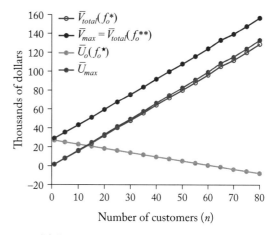

(b) Insurer's profit as function of n. The insurer does not gain by underwriting the SP's customers and attributing the loss to the SP.

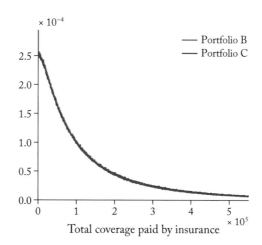

(c) Probability Distribution Function (pdf) of the amount paid out by the insurer in different scenarios.

Figure 4.9: Optimal incentive factor, insurer's profit, and pdf of the coverage paid by insurer under loss model (4.26).

4.4.2   SOCIAL WELFARE

Our analysis so far has focused on whether it is in the interest of an underwriter to insure risk-dependent insureds, and if so how best to do so. We now turn to the issue of social welfare, i.e., whether by embracing risk dependency the underwriter can also help improve the total utility. We have shown that underwriting both SP and his customers and giving SP more discount on premium improves the insurer profit and decreases the probability of data breach. As a consequence of the latter, the utility of the insureds improves; thus underwriting both SP and his customers indeed improves the social welfare (total utility) in general.

Let $C_o(f_o')$ and $C_i(f_o')$ be the total expected cost paid by the SP and his customer, respectively. We have,

$$C_o(f_o') = \underbrace{b_o f_o}_{\text{Premium}} + \underbrace{E[D_o] \cdot P_o(f_o - f_o')}_{\text{Expected uncovered loss}}$$

$$C_i(f_o') = \underbrace{b_i f_i}_{\text{Premium}} + \underbrace{E[D_i] \cdot P_{li}(f_o - f_o')}_{\text{Expected uncovered loss}}, \tag{4.28}$$

where $D_i = \begin{cases} L_i \text{ if } L_i \leq d_i \\ d_i \text{ o.w} \end{cases}$ is the amount of deductible that insured i pays. Note that we do not consider discount $b_o f_o'$ in the SP's costs because this is assumed to be used toward its security investment. The social welfare $SW(f_o')$ is the insurer's profit less the costs:

$$SW(f_o') = \overline{V}_{total}(f_o') - C_o(f_o') - \sum_{i=1}^{n} C_i(f_o'). \tag{4.29}$$

Below we use an example similar to that provided in Section 4.3.3 to illustrate the impact the insurance policy has on social welfare.

Consider an SP and a single customer, and assume that both have a large annual revenue ($10B–$100B), with a base rate $b_o = b_1 = \$52,000$ and base retention $d_o = d_1 = \$250,000$. We assume that $f_o = 1.2$, $f_1 = 1$, $P_o(f) = P_1(f) = \frac{0.05}{1+\frac{b_o(1.2-f)}{1000}}$, and $t = 0.5$. Based on Table 4.5, we assume both L_o and L_1 have a log-normal distribution with mean $\$5,965,571$ and median $\$3,326,313$.

We now compare two cases. In the first case the insurer ignores the risk dependency and attempts to separately maximize her profit from the SP and his customer, respectively. In the second case, the insurer jointly optimizes the two policies.

In the first case, the insurer obtains the discount to the SP as follows:

$$l_o = l_1 = E((L_o - d_o)^+) = \$5,715,600$$
$$\overline{V}_o(f_o') = b_o \cdot (f_o - f_o') - l_o P_o(f_o - f_o') \Rightarrow f_o^* = 0.3057. \tag{4.30}$$

The insurer's profit and the insureds' costs are as follows:

- Insurer's total expected revenue:

$$\overline{V}_{total}(f_o^*)$$
$$= b_o(f_o - f_o^*) - l_o P_o(f_o - f_o^*) + b_1 f_1 - P_{l1}(f_o - f_o^*)l_1$$
$$= 52000 \times (1.2 - 0.3057) - 5715600 \times 0.003 + 52000 - 5715600 \times 0.0059$$
$$= \$47,635$$

- SP's expected cost:

$$C_o(f_o^*) = b_o f_o^* + E[D_o]P_o(f_o - f_o^*) = 52000 \times 1.2 + 28753 \times 0.003 = \$62,486.$$

- SP's customer's expected cost:

$$C_1(f_o^*) = b_1 f_1 + E[D_1]P_{l1}(f_o - f_o^*) = 52000 \times 1 + 28753 \times 0.0059 = \$52,170.$$

- Total utility or social welfare:

$$SW(f_o^*) = \overline{V}_{total}(f_o^*) - C_o(f_o^*) - C_1(f_o^*)$$
$$= \$47,635 - \$62,486 - \$52,170 = -\$67,021.$$

In the second case, the insurer jointly maximizes policies for the SP and its customer. It obtains the optimal incentive factor as follows:

$$\overline{V}_{total}(f_o') = b_o(f_o - f_o') - l_o P_o(f_o - f_o') + b_1 f_1 - P_{l1}(f_o - f_o')l_1 \Rightarrow f_o^{**} = 0.3785.$$

The insurer's profit and the insureds' costs are given by:

- Insurer's total expected revenue:

$$\overline{V}_{total}(f_o^{**})$$
$$= b_o \cdot (f_o - f_o^{**}) - l_o P_o(f_o - f_o^{**}) + b_1 \cdot f_1 - P_{l1}(f_o - f_o^{**})l_1$$
$$= 52000 \times (1.2 - 0.3785) - 5715600 \times 0.0024 + 52000 - 5715600 \times 0.0056$$
$$= \$48,993.$$

- SP's expected cost:

$$C_o(f_o^{**}) = b_o \cdot f_o^{**} + E[D_o]P_o(f_o - f_o^{**})$$
$$= 52000 \times 1.2 + 28753 \times 0.0024 = \$62,469.$$

- SP's customer's expected cost:

$$C_1(f_o^{**}) = b_1 \cdot f_1 + E[D_1]P_{l1}(f_o - f_o^{**})$$
$$= 52000 \times 1 + 28753 \times 0.0056 = \$52,161.$$

- Total utility or social welfare:

$$SW(f_o^{**}) = V_{total}(f_o^{**}) - C_o(f_o^{**}) - C_1(f_o^{**})$$
$$= \$48,993 - \$62469 - \$52161 = -\$65,637.$$

We see that the total utility or social welfare is higher in the second case, when the insurer takes risk dependency into account and jointly optimizes the two policies. It is interesting to note that the values used in this example lead to negative social welfare, i.e., the total cost born by the insureds exceeds the total profit made by the insurer. The negative total utility is a reflection of the damage inflicted by attackers behind data breaches.

4.4.3 MODELING THIRD-PARTY LIABILITY

We have assumed that the probability that the insurer can attribute a part of the loss to the third party is a constant (q) and is independent of P_o, P_i, and t. An alternative model is to find probability q using P_o, P_i, and t. Let q_i be the probability that the insurer of insured i can attribute a part of the loss to its third party. Moreover, define events A_i and B_i as follows:

- A_i: a business interruption occurs to insured i due to a data breach/loss incident on the SP's side.

- B_i: a loss incident occurs to insured i.

We then have:

$$Pr\{A_i \cap B_i\} = P_o(f_o - f_o') \cdot (1 - P_i(f_i)),$$
$$P\{B_i\} = P_{li}(f_o - f_o', f_i) = P_i(f_i) + tP_o(f_o - f_o') - tP_o(f_o - f_o')P_i(f_i),$$
$$q_i = Pr\{A_i|B_i\} = \frac{Pr\{A_i \cap B_i\}}{Pr\{B_i\}} = \frac{P_o(f_o - f_o')(1 - P_i(f_i))}{P_i(f_i) + tP_o(f_o - f_o') - tP_o(f_o - f_o')P_i(f_i)}.$$

The above equation implies the assumption that the insurer is always able to attribute the loss of insured i to the SP if the latter is the cause of the loss. Under this assumption, Equations (4.18) and (4.19) can be written as follows:

$$\overline{U}_i(f_o') = b_i \frac{f_{\min} + f_{\max}}{2} - E[P_i(f_i)]l_i, \tag{4.31}$$

$$\overline{U}_o(f_o') = b_o(f_o - f_o') - P_o(f_o - f_o')\left[l_o + t\sum_{i=1}^{n}(1 - E[P_i(f_i)])l_i\right]. \tag{4.32}$$

These two equations are equivalent to Equations (4.18) and (4.19), respectively, by setting $q = 1$ in (4.18) and (4.19). Therefore, all the theorems continue to hold for $q_i = Pr\{A_i|B_i\}$.

Note that the third-party liability $t\sum_{i=1}^{n}(1 - E[P_i(f_i)])l_i$ may be large, in which case b_o would also be large, for otherwise insuring SP alone is not profitable for the insurer. If insuring the SP alone is not viable due to high third-party liability, then neither portfolio A nor C is viable, and portfolio B becomes the only choice.

4.4.4   NON-MONOPOLISTIC INSURER

Our analysis has assumed a monopolistic insurer. The modeling choice is aimed at focusing rather singularly on the issue of risk dependency without the interference of competition. Without monopoly the insurer will have to consider giving up her profit, but it does not change our main message. Our analysis simply points to the fact that if the insurer recognizes the risk dependency among the insureds, then with the right incentive she can extract more profit; without monopoly she might have to give up all of this profit.

Nonetheless, if there is competition, which often drives profit down to zero depending on the model, it may not be in the interest of the insurer to recognize this risk dependency or incentivize the SP. On the other hand, if one insurer is competing with another who is ignorant of the risk dependency among her prospective clients, then the first insurer now has an advantage in recognizing this and can effectively lower her cost of providing insurance and be able to offer more competitive contracts (with lower premium, i.e., returning a share of the profit to the insureds).

4.5   CHAPTER SUMMARY

This chapter follows a standard underwriting framework commonly used in the insurance industry, and shows what our earlier insights on insuring dependent risks translate into in this framework. In the previous chapter we showed that contrary to a common dependency-avoidance practice, there is an unrealized incentive for an insurer to underwrite dependent risks, as the existence of risk dependency among multiple insureds allows the insurer to jointly design polices that incentivize the insureds to (collectively) commit to higher levels of effort, which results in improved state of security for all and in improved profits for the insurer. In this chapter, by focusing more specifically on a one-way risk dependency, that between a service provider and his multiple customers, we see clearly that the insurer is able to achieve higher profit by insuring all of them provided she appropriately incentivizes the service provider to improve his state of security, which leads to risk reduction for all his customers, thus a benefit multiplier for the insurer.

4.6   TABLE OF NOTATIONS USED IN THIS CHAPTER

See Table 4.6.

Table 4.6: Notations used in the two-agent model (*Continues.*)

Symbol	Definition
$i = 0, 1, \ldots, n$	Indexing the agents purchasing insurance: $i = 0$ denotes the SP; $i = 1, 2, \ldots, n$ denote the SP's customer agent i
b_i	Base premium on i's policy
f_i	Modifier factor applied to i's base premium, uniformly distributed over some $[f_{min}, f_{max}]$
L_i, d_i	Agent i's (random) loss and his elected retention
f_o'	SP's improved security posture measured by decrease in assessed modifier factor value
$\tilde{f}_o$	SP's modifier factor after improvement/incentives
$b_i f_o'$	Discount in premium given to the SP
$P_o(\tilde{f}_o)$	Probability of a breach to the SP
$P_i(f_i), \overline{P}_i(f_i)$	Probability of a loss incident to agent i unrelated to the SP, and its expectation when f_i is randomly generated
t	Probability of business interruption to i as a result of breach to SP
$P_{li}(\tilde{f}_o, f_i), \overline{P}_{li}(\tilde{f}_o, f_i)$	Total probability of a loss event to i, and its expectation when f_i is randomly generated
$V_o(f_o'), \overline{V}_o(f_o')$	Insurer's utility/profit and expected utility/profit from underwriting only the SP
I_o	Bernoulli random variable with parameter $P_o(f_o - f_o')$
$V_i(f_o'), \overline{V}_i(f_o')$	Insurer's utility from underwriting i
$I_i, i = 1, \ldots, n$	Bernoulli random variable with parameter $P_{li}(f_o - f_o', f_i)$
$l_i = E[(L_i - d_i)^+], i = 0, \ldots, n$	Agent i's expected excess loss over deductible, introduced for convenience
$\overline{V}_{total}(f_o'), \overline{V}_{max}$	Insurer's utility when underwriting both the SP and his n customers, and when maximized over the choice of f_o'
f_o^*, f_o^{**}	Optimizers of $\overline{V}_o()$ and $\overline{V}_{total}()$, respectively

Table 4.6: (*Continued.*) Notations used in the two-agent model

Symbol	Definition
$W, g()$	The noise and its pdf in the assessed security modifier factor
$\Pi_o(f_o - f'_o) := E[P_o(f_o - f'_o + W)]$	Introduced for convenience
q	Probably that a loss can be attributed to the SP by his customer
$U_i(f'_o), \overline{U}_i(f'_o)$	Insurer's utility and expected utility under Portfolio C, only underwriting the SP's customers, as a function of the incentive given to the SP (by the third-party insurer)
J_i	Bernoulli random variable with parameter $P_i(f_i) + (1-q)(tP_o(f_o - f'_o)(1 - P_i(f_i)))$
$U_o(f'_o), \overline{U}_i(f'_o)$	Third-party insurer's utility and expected utility in underwriting the SP and offering incentive f'_o
K_i	Bernoulli random variable with parameter $q(tP_o(f_o - f'_o))(1 - P_i(f_i))$
$f_o^\star$	The maximizer of $\overline{U}_o(f'_o)$; this is the optimal level of incentive offered by the SP's insurer to the SP
$\overline{U}_{max} := \sum_{i=1}^{n} \overline{U}_i(f_o^\star)$	Insurer's maximum utility under Portfolio C

4.7 APPENDIX

4.7.1 PROOF OF THEOREM 4.3

Recall from (4.21) that for $0 \le f'_o \le f_o$ we have $\overline{U}_o(f'_o) + \sum_{i=1}^{n} \overline{U}_i(f'_o) = \overline{V}_o(f'_o) + \sum_{i=1}^{n} \overline{V}_i(f'_o) = \overline{V}_{total}(f'_o)$. We will assume that $\overline{U}_o(f_o^\star) \ge 0$, for otherwise no insurer will underwrite the SP. Since $f_o^{\star\star}$ is the optimizer of $\overline{V}_{total}(f'_o)$, we have

$$
\begin{aligned}
\overline{V}_{\max} = \overline{V}_{total}(f_o^{\star\star}) \; &\ge \; \overline{V}_{total}(f_o^\star) \\
&= \; \overline{V}_o(f_o^\star) + \sum_{i=1}^{n} \overline{V}_i(f_o^\star) \\
&= \; \overline{U}_o(f_o^\star) + \sum_{i=1}^{n} \overline{U}_i(f_o^\star) \ge \overline{U}_{\max}. \qquad (4.33)
\end{aligned}
$$

Similar to the proof of Theorem 4.1, by the first-order condition we have

$$f_o^{\star} = \left(f_o - (P_o')^{-1} \left(\frac{b_o}{l_o + q \sum_{i=1}^n l_i t (1 - \overline{P}_i(f_i))} \right) \right)^+ . \tag{4.34}$$

From the proof of Theorem 4.1, we have

$$f_o^{\star\star} = \left(f_o - (P_o')^{-1} \left(\frac{b_o}{l_o + \sum_{i=1}^n l_i t (1 - \overline{P}_i(f_i))} \right) \right)^+ . \tag{4.35}$$

Since $P_i'(.)$ is an increasing function and

$$\frac{b_o}{l_o + q \sum_{i=1}^n l_i t (1 - \overline{P}_i(f_i))} > \frac{b_o}{l_o + \sum_{i=1}^n l_i t (1 - \overline{P}_i(f_i))}, \tag{4.36}$$

we have $f_o^{\star} \le f_o^{\star\star}$.

CHAPTER 5

How to Pre-Screen: Risk Assessment Using Data Analytics

The discussion of previous chapters invokes the critical notion of pre-screening and relies on it heavily. Indeed, our principal message has been that insurance can be used as an effective incentive mechanism, provided that the insurer has the ability to pre-screen, i.e., reasonably accurately assess the "effort" exerted by the insured, or equivalent the insured's cyber risk. In practice, this information is obtained through various proxies.

The first is by using questionnaires as mentioned in Chapter 4: survey forms that a prospective agent is asked to fill out when applying for an insurance policy. We saw examples of the types of questions in Chapter 4; the answers to these questions determine the premium through a sequence of table lookups.

The second is by using methods such as audits and penetration tests. This typically involves onsite, close-up examinations of the prospective agent's software systems, file systems, data storage, access control, etc. The output of the audit can either be used to directly determine the premium as in the previous case, or be fed as input to a model or formula that generates risk estimates.

In this chapter, we present an approach that combines large-scale Internet measurement and predictive analytics, which automates the process of cyber risk assessment and quantification and allows it to be done at scale. This assessment approach directly estimates an organization's risk of having a material cyber incident such as a data breach, by using a supervised learning technique, in a domain where machine learning technologies are just starting to be recognized as a potentially powerful tool.

Central to our research is the question "*to what extent can we predict data breaches and other cybersecurity incidents for an organization?*" The ability to do so has far-reaching social and economic impact beyond its use in cyber insurance. Often, by the time a breach is detected, it is too late with damages already done. Consequently, being able to predict such incidents can greatly enhance an organization's ability to put preventative measures in place and make much more judicious resource allocation decisions in doing so.

The work presented in this chapter draws from our publications [84, 103, 126]. These prototype studies tap into a fairly wide variety of Internet measurement data related to a network's malicious activities, mismanagement, misconfiguration, and latent threats, including a large number of host reputation blacklists that we detail below.

5.1   PREDICTIVE POWER OF MEASUREMENT DATA

We will start by taking a look at some examples of data and see what we can do with them. Our general idea is to understand to what extent these data can help us tell apart two types of organizations: those who have had a reported data breach in the recent past (referred to as the victims or the victim group/set) and those who have not (referred to as non-victims or the non-victim group/set). To do so, the first thing we need is data breach information.

Cyber incident/breach data.   When we first did this type of study in 2014, cyber incident reporting was very uneven: there was no standard governing the format of incident disclosure. Reporting was largely driven by regulatory requirements, which varied (and continue to vary) from region to region. As a whole, cyber incidents were very under-reported. The situation has improved over the past five years, with more and larger data repositories now in existence, some of which run by commercial entities serving the insurance industry. The studies reported in this chapter relied on some of the best data sources that were publicly available then and detailed below.

- *VERIS Community Database (VCDB)* [120], a collection maintained by the Verizon RISK Team and used by Verizon in its highly publicized annual Data Breach Investigations Reports (DBIR) [121],

- *Hackmageddon* [97], an independently maintained cyber incident blog that aggregated and documented various public reports of cyber security incidents on a monthly basis, and

- the *Web Hacking Incidents Database (WHID)* [40], an actively maintained cybersecurity incident repository, with a goal to raise awareness of cybersecurity issues and to provide information for statistical analysis.

Table 5.1 summarizes the main categories of incidents contained in the three datasets we used and their volume.

Mismanagement Data
We now introduce a first type of measurement data that pertain to how a network system, including its servers and devices, is configured and managed. These measurements typically capture deviations from standards, known best practices, and other operational recommendations. Some examples are given below.

- *Untrusted HTTPS Certificates*: these refer to web servers/websites not configured to present a valid certificate (signed by a trusted authority) during the Transport Layer

Table 5.1: Reported cyber incidents by category. Only the major categories in each set are shown. The "Else" category by VCDB represents incidents lacking sufficient detail for better classification.

Incident Type	SQLi	Hijacking	Defacement	DDoS
Hackmageddon	38	9	97	59
WHID	12	5	16	45
Incident Type	Crimeware	Cyber Esp.	Web App.	Else
VCDB	59	16	368	213

Security (TLS) handshake [42]. Sometimes one was never obtained; more often a valid certificate has expired. Malicious websites (e.g., a phishing site) do not have valid certificates in general, so the absence of a valid certificate would have been a very effective way of identifying such sites; unfortunately, too many benign sites are not configured correctly, effectively rendering the certificate's absence a far less useful signal in practice.

- *Open Recursive Resolvers*: these refer to misconfigured open Domain Name Server (DNS) resolvers [16, 41], which can be easily used to facilitate massive amplification attacks that target others. In short, an attacker can send simple DNS queries to an open resolver with a spoofed source IP address (the victim's), resulting in overwhelming DNS responses to the victim. As a rule, recursive lookups should be disabled unless specifically required and, when enabled, limited to intended customers.

- *DNS Source Port Randomization*: this refers to DNS servers not patched to implement source port randomization as recommended by RFC 5452 [60] to minimize the threat of DNS cache poisoning attacks [13].

In each of these cases, we simply calculate the percentage of misconfigured (or symptomatic) servers of all such servers within an entity[1] and plot the cumulative distribution function (CDF) of these fractions for the victim and non-victim populations as defined by the VCDB breach dataset. These are shown in Figure 5.1. We see a clear difference between the two populations in how their untrusted HTTPS and Open Resolvers are distributed. This difference suggests that these two measurements are meaningful distinguishers, and thus hold predictive power. This is indeed verified later when these two symptoms emerge as the most significant among a set of candidates we experimented with. By contrast, the two populations do not show significant difference in their distribution of the third measurement, the DNS port randomization.

---

[1]This step is tedious but highly non-trivial, and requires the use of IP intelligence data [15] that summarize organizational boundaries in IP blocks.

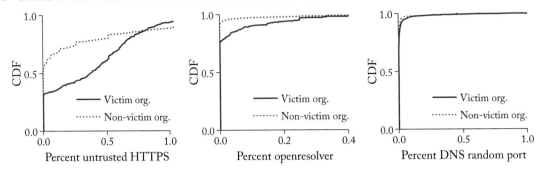

Figure 5.1: Comparison of mismanagement symptoms between the victim and non-victim populations. There is a clear separation under the first two, while the third appears to be a much weaker predictor.

It is worth pointing out that the above type of measurements identify a network's misconfigurations, which are not vulnerabilities in and by themselves. The presence of misconfigurations in an organization's networks and infrastructure is, however, an indicator of a lack of appropriate policies and/or technological solutions to detect such failures. This lacking does increase the potential for a successful data breach.

While these measurements were taken during the same timeframe covered by the breach data, we are not being very strict about their exact timing in the above illustrations (e.g., whether measurements on victims were taken before or after their data breaches). We will be much more precise and restrictive when we perform the actual prediction exercise presented later in this chapter.

Malicious activity metrics.   We can similarly examine a second type of measurements that pertain to explicit signs of infections in a network system and its devices, in particular, botnet activities [112]. These are referred to as host-level malicious activities, typically captured by well-established monitoring systems such as spam traps [21], darknet monitors [8], or DNS monitors. These observations are then distilled into blacklists, called host reputation blacklists (RBLs) [4]; each blacklist is refreshed on a daily basis and consists of a set of IP addresses seen to be engaged in some malicious activity. These measurements are signs of compromises already in a system (machines taken over for nefarious activities), though they are not considered data breaches (as they may not involve data losses) in the sense we mean. For this reason, we will not call these breaches. These RBLs further breaks down into three types:

1. those capturing spam activities, including CBL [5], SBL [24], SpamCop [20], WPBL [27], and UCEPROTECT [26];

2. those capturing phishing and malware activities, including SURBL [23], PhishTank [19], and hpHosts [12]; and

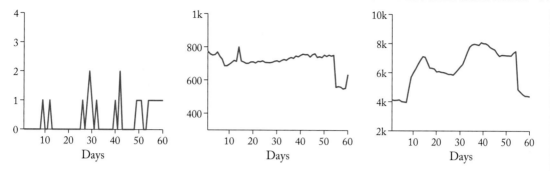

Figure 5.2: Malicious activity time series of three organizations: Org 1, Org 2, and Org 3, respectively.

3. those capturing scanning activities, including the Darknet scanners list, Dshield [10], and OpenBL [17].

To see whether we can use these measurements to tell apart the victims from the non-victims, we again first need to aggregate this host-level data to the firm/entity level and we do so simply by counting the daily total of an entity's unique IPs on these lists (all or a subset of these lists).[2] This aggregation results in one or more time series for an entity, encoding its presence on these blacklists at a sample rate of once a day. Since we are now dealing with time series, questions arise as to what kind of features or quantities to extract for our purpose (of telling apart the two populations). It turns out we now have an opportunity to capture distinct behavioral patterns in an organization's malicious activities, particularly concerning their dynamic changes.

To illustrate, the three examples given in Figure 5.2 show drastically different behaviors: Org 1 shows a network with consistently low levels of observed malicious IPs (and possibly within the noise inherent in the blacklists), while Orgs 2 and 3 show much higher levels of activity in general. These two, however, differ in how persistent they are at those high levels. Org 2 shows a network with high levels throughout this period, while Org 3 shows a network with more fluctuation. Intuitively, such dynamic behavior reflects to a large degree how responsive the network operators are to blacklisting, such as time to clean up, time to resurfacing of malicious activities, and so on.

These observed differences motivate us to collect statistics summarizing such behavioral patterns by measuring their persistence and change, e.g., how big is the change in the magnitude of malicious activities over time and how frequently does it change. To balance the expressiveness of the features and their complexity, we first value-quantize a time series into three regions relative to its time average: "good," "normal," and "bad."

---

[2]This step again requires the use of IP intelligence data [15] that allows us to associate individual IP addresses with given organizations.

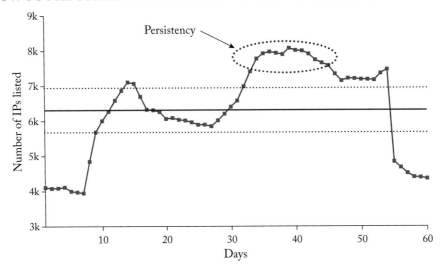

Figure 5.3: Extracting temporal measures.

An illustration is given in Figure 5.3 using one of the examples shown earlier (Org 3). The solid line marks the average magnitude of the time series over the observation period; the dotted lines then outline the "normal" region, i.e., a range of magnitude values that are relatively close (either from above or below) to its time-average. The region above the top dotted line is accordingly referred to as the "bad" region, showing large number of malicious IPs, and the region below the bottom dotted line the "good" region, with a smaller number of malicious IPs, both relative to its average. Note that this quantization step also serves to capture certain onset and departure of "events," such as a wide-area infection, or scheduled patching and software update, etc. Viewed this way, the duration an organization spends in a "bad" region could be indicative of the delay in responding to an event, and similarly, how frequent it re-enters a "bad" region could be indicative of the effectiveness of the actions taken to remain clean. For each region we then measure the average magnitude (both normalized by the total number of IPs in an organization and unnormalized), the average duration that the time series persists in that region upon each entry (in days), and the frequency at which the time series enters that region.

We are now ready to see whether these measures have the power to separate the two populations, by comparing their distributions among the victims and non-victims, respectively. Here we count the number of unique IP addresses listed on a given day by the set of scanning blacklists that belong to an entity. This step is carried out in the same way for both victim and non-victim organizations. Figure 5.4 shows this comparison for four examples. We see that in each case there is a clear difference between the two populations in how these feature values are distributed, e.g., victim organizations tend to have longer bad periods, indicative of slow response time, and also higher bad/good magnitudes, etc.

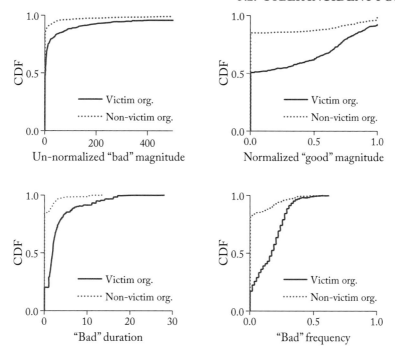

Figure 5.4: Example temporal measures from a time series over a 30-day period.

5.2   CYBER INCIDENT FORECAST

We next describe a security incident forecasting framework we developed and detailed in [84]. It has two ingredients: (1) the utilization of the types of measurement data shown earlier, that collectively characterize the security posture of organizations, as well as a set of carefully extracted features from these datasets, examples of which are presented in the previous section; and (2) the use of security incident reports to determine the security outcomes of organizations.

In the language of supervised learning [22], an area of machine learning, the first ingredient constitutes the *feature* information, the second the *label* information; for a given organization, its combination of feature and label constitutes an input-output pair. The learning task is to use a sufficient number of known input-output pairs to train a classifier, which is then able to take a given input and determine the most likely output. In out context, this means to use both victim organizations (their features and the fact they are victims) and non-victim organizations (their features and the fact they are non-victims) to train the classifier, which has the ability, when presented with a set of features, to determine whether these features are more likely to belong to a victim or a non-victim. It is also important to note that our goal is for the classifier to be a *predictor*: to be able to *predict* the likelihood of a future cyber incident for a given set of features. This means we need to train the classifier with feature and label pairs that are carefully aligned

Table 5.2: Chronological separation between training and testing samples for each incident dataset; the split is roughly 50–50 among the victim population

	Hackmageddon	VCDB	WHID
Training	Oct 13–Dec 13	Aug 13–Dec 13	Jan 14–Mar 14
Testing	Jan 14–Feb 14	Jan 14–Dec 14	Apr 14–Nov 14

in time, with feature information strictly predating the label information (measurements taken prior to the known breach). We elaborate on this more below.

Feature sets and training procedure.   We use two types of features: a primary set and a secondary set. The primary set of features consists of the raw data, while the secondary set is derived or extracted from the raw data, i.e., in the form of various statistics, some of which were discussed in detail in the previous section.

The training step uses two sets of samples (input-output pairs), a subset of victim organizations (victim group) and a randomly selected set of non-victim organizations (with size comparable to that of the victim group in any given experiment). The victim group is selected from our incident dataset based the time stamps of the reported incidents: we use those occurred earlier for training and those occurred later for testing, with a typical split ratio of 50–50. This random selection of non-victim organizations is repeated numerous times, each time training a different classifier. The reported testing results are averages over all these versions.

Using this methodology, we performed prediction using the three incident datasets separately, as well as collectively. When used collectively, we removed duplicate reports of the same incident whenever applicable. The separation between training and testing for each dataset is done chronologically, so that it results in an approximate 50–50 split of the victim set between the training and testing sample sizes.

When training a classifier using non-sequential data (e.g., pictures of cats and dogs), the split of samples for the purpose of training and testing is often done randomly in the machine learning literature. Due to the sequential nature of our data, we intentionally and strictly split the data by time: earlier ones are for training and later ones for testing. Because of this, our testing results are indeed "prediction" results.

Also worth noting is that feature information of a victim organization used for training is collected over a period of time prior to the month in which its incident occurred.[3] As an illustration, Figure 5.5 shows the use of a 2-month period (referred to as Recent-60 features) and the use of a 2-week period (referred to as Recent-14 features). For a non-victim used for training the features are again collected over a similar period prior to the month of the earliest incident in the victim group used for training. The features used for testing are obtained in a

---

[3]Most incident occurrences in our dataset are timestamped with month and year information.

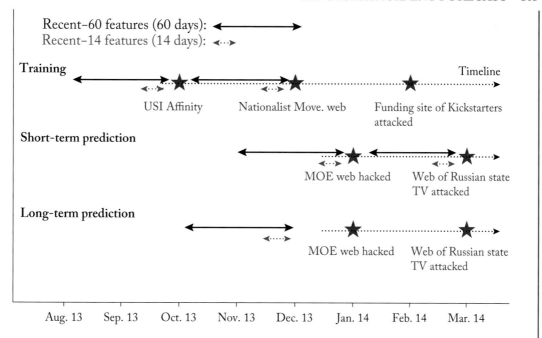

Figure 5.5: Feature extraction and short-term and long-term forecasting. Red stars denote specific incidents. In training, features are extracted from the most recent period leading up to an incident. In testing, the same is done when we perform short-term forecast; in long-term forecast, features are extracted from periods leading up to the time of the first incident used in testing.

similar way for the victims and non-victims with the following distinction: when victim features are collected prior to each incident, this essentially allows us to assess the *short-term forecast* capability of the model, whereas if we collect these features prior to the earliest incident of the group, then this allows us to assess the *long-term forecast* capability of the model since an incident may occur months after the features are collected (up to 12 months in the case of VCDB). This is also illustrated in Figure 5.5.

Prediction result. The type of classifier we trained is a set of decision trees following the Random Forest (RF) method [83], which is an ensemble classifier and an enhancement to the classical random decision tree method. The output of the classifier is a risk probability; a threshold is then imposed to obtain a binary label. For instance, if we set the threshold at 0.5, then all output > 0.5 means a label of 1. The accuracy of the prediction can then be quantified by comparing the binary prediction after thresholding against the ground-truth: a victim has a true label of 1 while a non-victim has a true label of 0. A true-positive (TP) probability is obtained by calculating the % of true labels of "1" correctly predicted, and a false-positive (FP) obtained

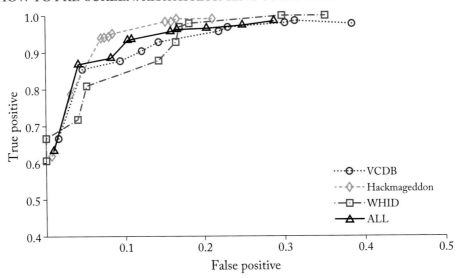

Figure 5.6: Prediction results.

by the % of true labels of "0" falsely predicted as "1". The TP-FP pair changes as we move the threshold; in general as one improves the other deteriorates. By moving this threshold from 0 to 1, we obtain different combinations of TP and FP, or different operating points of the prediction performance. The spectrum of these combinations constitute a ROC (receiver operating characteristic) curve when plotting TP against FP. The larger the area under the curve (AUC), the better the predictor. Operationally, a good threshold choice is where there is a clear "knee" in the curve.

The prediction results (for a 50–50 split of the victim set between training and testing, and in the long-term forecast test setting) are summarized in the set of ROC curves shown in Figure 5.6. Since our non-victim set for training is randomly selected from the (much larger) total non-victim population, the above test is repeated multiple times for a given threshold value, each time for a different random non-victim set. The average TP and FP over these repeated tests form one point on the ROC curve. We see the prediction performance varies slightly between the datasets, but remains very satisfactory, generally achieving combined (TP, FP) values of (90%, 10%) or (80%, 5%). In particular, when we combine the three datasets, we can achieve an accuracy level of (88%, 4%). A summary of some of the most desirable operating points are given in Table 5.3.

Top data breaches circa 2015.   We also plot the prediction output without thresholding for the VCDB victims and a random non-victim set from 2015–2016 in Figure 5.7, and highlight some of the major data breach victims of that year. Clearly, this type of predictive analytics can enable proactive measures by an organization as well as enable more accurate and quantitative

Table 5.3: Best operating points of the classifier for the best combinations of (TP, FP) values

Accuracy	Hackmageddon	VCDB	WHID	All
True positive (TP)	96%	88%	80%	90%
False positive (FP)	10%	10%	5%	10%
False negative (FN)	4%	12%	20%	10%
Overall accuracy	90%	90%	95%	90%

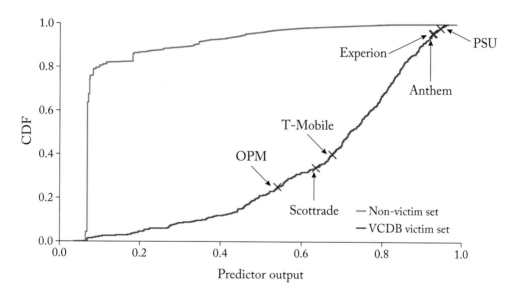

Figure 5.7: Top data breaches between 2015 and 2016. (OPM: Office of Personnel Management; PSU: Penn State University.)

cyber risk assessment in vendor/third-party evaluation and the design of better cyber insurance policies.

5.3   FINE-GRAINED PREDICTION

In a second study [103], we set out to perform a more fine-grained prediction which produces not the overall risk estimate but a *risk profile* that estimates the probabilities of an entity falling victim to each of a variety of security incidents *conditioned* on an incident occurring. This conditional probability estimate can then be combined with the study shown in the previous section to generate both the overall risk estimate and the more detailed risk profile estimates.

Our motivation behind this study is the desire to estimate the distribution of risk among multiple incident types, which then allows us to narrow down the recommendation on the most effective preventive measures. Depending on the types of incidents one is most likely to face, a firm can make much more judicious and strategic resource allocation decisions. The ability to pinpoint the nature of the estimated risks, not just the overall level of the risk, can greatly enhance the usability and actionability of the risk quantification. For instance, a large healthcare organization and a large IT company may have similar levels of overall risk, but the former may be much more subjective to malware and hacking while the latter thefts. To distinguish these different risks and estimate them separately can provide very valuable information for an organization in its strategic decision-making and resource allocation. Similarly, distribution of risk among different types of data incidents can also help insurance providers better assess the potential amount of loss which in turn helps determine the contract terms, including premiums and coverage levels.

The previous study focuses on capturing various aspects of an organization's security posture, and in particular focuses on capturing features that tend to be stable over a long period of time (e.g., on the order of months). While this has proven effective in predicting the overall risk an organization faces, to obtain fine-grained prediction that estimates risks of specific type of incidents we also tapped into a richer set of data, which includes the following.

- Business sector information, which may be more relevant in capturing highly targeted attacks; there has been evidence suggesting, see, e.g., [114], correlation between business/employee risk factors and one's likelihood of becoming the target of a (spear) phishing attack.

- IP Intelligence data obtained from Neustar [15] that maps IP address blocks to their respective business owners and their industry sectors, and

- Various web statistics gathered from Alexa Web Information Services (AWIS) [29].

Together these data sources allow us to ascertain business features of an organization (region of operation, business sector, employee count, etc.), their web statistics (ranking, traffic volume, number of visits, speed, number of pages linking to the website, popularity, etc.), and determine the network assets or cyber footprint of a given domain.

A cascade of binary classifiers. To provide prediction for specific incident types brings into even sharper focus the deficiencies in the state of incident reporting. Besides under-reporting, even among those reported incidents there is a general lack of uniformity and standard terminology as to create difficulty in discerning exactly what happened. This type of reporting may still be useful in the framework presented in the previous section, as a description of "something happened"[4] is sufficient to provide us with a label, making it a legitimate data point. However, if we need to know what type of incident occurred, this type of description is grossly insufficient. This motivates us to seek techniques to mitigate the deficiencies in reporting and to augment

---

[4]This is an extreme example but can be found in datasets like VCDB.

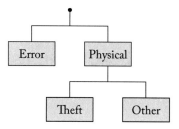

Figure 5.8: Risk assessment tree.

our prediction framework to generate credible fine-grained prediction of incident types as we describe below.

In pursuing more fine-grained prediction, we need to take into account more detailed descriptions of an incident or the incident signature (e.g., action, actor, and asset involved in an incident [121]). A naive way is to take the entire incident signature of an incident as a class label, and the victim features as input data for the classifier. However, given the large number of possible incident signatures (different combinations of action, actor, asset, among others), there would only be a small number of samples per signature vector (a particular combination). Furthermore, a significant number of incident entries provide only partial information about their corresponding incident as mentioned earlier. Ignoring such entries will leave us with even fewer samples. Thus, such a straightforward application of classifier construction is clearly infeasible in practice.

Our solution to this problem is to build multiple classifiers, each of them estimating a portion of the incident signature. An interpretation of this technique is the chain rule in probability. To illustrate, consider estimating the risk factor for an organization of type t to experience a physical theft incident. We can break this risk into two parts: $Pr(\text{Physical Theft} \mid t) = Pr(\text{Physical} \mid t) \, Pr(\text{Theft} \mid \text{Physical}, t)$. As a result, entries that cite a physical incident without specifying additional details can still be included for building and testing the first classifier (i.e., physical, first term on the RHS), but will be ignored when building the second classifier (i.e., theft). This can be visualized as a tree shown in Figure 5.8, where each node represents a data breach type with increasing granularity further down the tree. The risk score at a node is the result of multiplying the risk at its parent node by the output of the classifier corresponding to said (child) node. Furthermore, if the number of samples in a given node is large enough, we can extend the tree by adding child nodes, providing us with more details about the incident signature. The classifier associated with each child node will answer a binary question about details of the incident, and add to the information gathered up to its parent node.

Fine-grained prediction result. Some examples of this fine-grained prediction are given in Table 5.4, where shaded cells are incident types deemed likely by our classifier (using a threshold) and those with crosses indicate actual incidents in the report. The results shown here highlight a

Table 5.4: Risk profiles for different sample organizations, and their corresponding industries' profiles

Organization	Error	Hacking			Malware	Misuse	Physical			Social
		Compromised Credential	Other				Theft	Other		
Information										
Russian Radio			×							
Verizon			×							
Public Admin.										
Macon-Bibb County	×									
The IRS						×				

main difference between the type of correlational studies done before, see, e.g., DBIR [121] that provides risk distribution for entire sectors ("Information" and "Public Administration" shown in the table), and the fine-grained prediction we seek to provide (for specific entities like "Verizon" and "IRS" in the table).

As shown in Table 5.4, information as a section is particularly at risk of hacking, malware and social engineering, but entities within the same sector may face very different risk profiles (more spread out in the case of Verizon than the sector average). Similarly, even though Public Administration as a sector is at risk of almost all types of incidents, the IRS is singularly prone to error and misuse (more concentrated risk than the sector average). Our framework thus allows us to generate more targeted prediction for individual organizations that could be far more relevant than the risk faced by the entire sector, which in turn enables more effective and proactive actions.

We estimate that on average an organization can protect against 90% of all incidents by focusing on < 70% of incident types; in some cases, the latter can be significantly smaller. An immediate consequence of this is that security investment and resource allocation decisions informed by such analysis are much more targeted and effective; thus this type of fine-grained prediction can be extremely useful for organizational strategic decision-making and prioritization.

5.4 BRINGING TECHNOLOGY TO MARKET

The technology detailed above collects on a global scale a wide variety of externally observed data on Internet organizations, and uses advanced data analytics to

1. generate a set of metrics on organizational security posture to enable *quantitative assessment of security posture*, and

2. use such assessment for the purpose of *forecasting cybersecurity incident*.

Our technology is unique because it was one of the first based on prediction, not detection. It targets organizations as a whole and not just its computers.

This brings about a paradigm shift by introducing new ways of thinking about cybersecurity, and new policies and mechanisms for network security, at a much higher level and in a more holistic manner. Indeed, there are many security policies and design decisions that can benefit enormously from such holistic assessment; some can only be meaningfully applied at a network or organizational level. The first category is incentive mechanisms, aimed at encouraging better security practices and investment by organizations. Cyber insurance is a prime example of this as the first few chapters of this book demonstrate, and it can only be applied at an organizational/network level. A second category include peering arrangements [18] and traffic routing decisions, where decisions have to be made based on an overall assessment of an entire network, not its many individual hosts. A third category involves security practices under severe resource limitation, e.g., deep packet inspection [9], whereby network-level assessment can enable a hierarchical and sequential form of inspection by allocating more resources in examining traffic from more malicious networks. Consequently, providing a sound quantitative framework for network-level security assessment is not only a scientific and logical choice, but also vital to a myriad of policy-making issues.

What has also become clear is that this risk quantification technology captured not only the cybersecurity conditions of individual hosts residing in an organization, but also various human elements, including cybersecurity spending, adequacy of the IT team, internal policies governing data access, sharing, procurement, and so on, that have direct implications on the organization's security outlook. These were generally not directly observable from the outside, but many of our measurements proved highly correlated with them. This is one of the key reasons why our technology worked—security is fundamentally as much a human problem as it is a technological problem.

Much of this technology has been transitioned to practice. It led to the world's first global enterprise cybersecurity ratings systems, commercialized through the startup company Quad-Metrics, Inc. I co-founded in 2014. As individuals, we all have a consumer credit rating that captures our individual default/delinquency rate. It is used whenever we apply for a credit card, a mortgage or other types of loan. What we built was essentially the analog of this credit rating but for organizations that captures their risk of suffering a material data breach. Indeed, most of us have a FICO consumer credit score issued by the analytics firm Fair Isaac (FICO). Our technology became the key ingredient underlying a parallel product by FICO called enterprise security score, when it acquired our startup in 2016.

The development of this technology and demonstrating that accurate cyber risk quantification can be done was critical in allowing us to frame the cyber insurance problem. As the previous chapters showed, our contract design framework heavily relies on this premise.

There were three primary market opportunities we identified when we first developed the technology between 2014 and 2016: (i) cyber insurance, (ii) third-party/vendor management, and (iii) enterprise security. Cyber insurance was the focus of work presented in this book and will not be repeated here. But both (ii) and (iii) are just as significant market needs.

Third-party/vendor management.   According to a 2013 Gartner report, "nearly 60% of IT budgets are spent on vendors for IT software, hardware, and services" [51]. With increasing regulation and high-profile breaches from vendors (e.g., both the Target and JP Morgan breaches in 2014 were third-party breaches as mentioned earlier), enterprise network managers are trying hard to understand and seeking ways to assess and monitor the risk and security posture of their vendors and partners whose networks they do not manage. Enterprises with franchises or that serve as a clearinghouse for financial transactions (brokerages) are particularly vulnerable. As in the case with insurance carriers, our quantitative risk assessment technology has found application in this space and filled this market need in vendor validation and management, allowing enterprises to make more informed decisions on vendor selection, and to change policies or take actions against risky vendors or partners.

Enterprise network security.   Worldwide information security spending is Information security spending is expected to reach $123.8 billion in 2020 [52], with roughly a third of that in the U.S. Despite significant spending, it is not always clear to network security professionals how effective their investment has been to reduce their current and future risk. For instance, complex mitigation devices can actually make things worse, if they are not understood or configured correctly, both of which may require costly personnel training. Even with this amount of security spending, the reality is that there are thousands of data breaches per year for the past 5–6 years [122]. At the same time, security threats are highly dynamic and evolves faster than counter measures, forcing IT security personnel to be constantly in a fire fighting mode, always a step behind; consider, e.g., heartbleed. There is thus an urgent need for a holistic and *proactive* approach to managing enterprise security risks. Our technologies fulfill this need by providing high visibility into issues and vulnerabilities within an organization and measuring security posture in a quantifiable way, which is the net result of all technical means, policies and procedures, and effectiveness of people implementing them. This provides IT security managers with the analytics to determine the effectiveness of security investments and policy and prioritize resources.

Becoming part of ESG: environmental, social, and corporate governance.   Much of the above market analysis remains true today. But cyber risk is also now emerging as an important measure of corporate governance that (institutional) investors increasingly pay attention to. ESG, or Environmental, Social, and Corporate Governance [11], refers to the leading factors in measuring the sustainability and societal impact of investment in a firm. The broader concept of ESG has been around for a few decades. Given potentially calamitous impact of climate change and widespread focus on sustainability, investors are increasingly look to factor sustainability issues into their investment choices, such as a firm's greenhouse gas emissions, carbon footprint, biodiversity, waste management, water management, and so on.[5] Similarly, social concerns include a firm's stance and/or record on diversity, human rights, consumer protection, and animal welfare; corporate governance concerns a firm's management structure and executive and employee compensation.

Within this context, there is rising interest by (institutional) investors in using cyber risk quantification as an additional key factor in measuring corporate responsibility, sustainability and societal impact. Given the direct relationship between data breaches and consumer privacy protection, cyber risk should indeed be recognized as a measure of corporate governance. Interestingly, the negative externality from poor cybersecurity is not unlike the negative externality from carbon emission from a variety of economic activities. Efforts to curb carbon emission and reduce cyber risks are both considered public goods and reduce the social cost to the public. The technology we developed is now being incorporated into the Institutional Shareholder Services' ESG ratings.[6]

It is possible that cyber may eventually become one of the four pillars in evaluating responsible corporate citizenship, ESCG: Environmental, Social, Cyber, and Corporate Governance.

5.5   CHAPTER SUMMARY

This chapter presents prototype studies on using measurement data to perform predictions of data breaches and other cyber incidents. In the first study, we primarily use two types of data: measurements on a network's misconfigurations or deviations from standards and other operational recommendations, and measurements on malicious activities seen to originate from that network, such as spam, malware, and phishing. The combination of both of these datasets represents a fairly comprehensive view of an organization's externally discernible security posture, and leads to very accurate prediction results. In the second study we tap into a richer set of data capturing information on business details to estimate firms' risk profiles that are much more targeted for individual organizations that are more meaningful and more actionable.

It is worth noting that (1) we focus exclusively on output measures, i.e., we collected data seen from the outside world, and not internal to a network, and (2) our datasets are very diverse and capture different aspects of an organization, ranging from the *explicit* or *behavioral*, such as externally observed malicious activities originating from a network (e.g., spam and phishing), to the *latent* or *relational*, such as mismanagement and misconfigurations in a network that deviate from known best practices, to the *structural*, such as an organization's business type, employee count and revenue. This diversity is crucial to our ability to predict not only cybersecurity incidents, but non-cyber security incidents, and those of a potentially unknown nature.

---

[5]My academic institution, the University of Michigan, announced in March 2021 that it would disinvest its multibillion-dollar endowment from fossil fuels as part of a push to have a net zero carbon investment portfolio, becoming the first public university to adopt such a pledge [25].

[6]ISS is a leading provider of ESG solutions and proxy voting for asset owners, asset managers, hedge funds, and asset servicing providers.

CHAPTER 6

Open Problems and Closing Thoughts

The driving force behind the rapidly increasing market demand for cyber insurance is the ever-present and ever-increasing severity of cyber threats and crimes, and very often the extremely high cost they incur not just in loss in revenue but also in terms of damage to a brand name. This book highlights two key elements in effective cyber insurance underwriting if the goal is not only risk transfer but also risk reduction: the ability to accurately assess an insured's cyber risk, and the ability to take into account risk dependencies. Over the last 4–5 years, the industry has steadily increased its focus on the former and is embracing new technologies; we presented a data driven predictive analytics framework to perform cyber risk quantification in Chapter 5. On the other hand, as illustrated in this book, standard insurance practice is not structured to readily take into account external risk dependencies, and this in turn is preventing the use of insurance policies as a more effective risk control mechanism.

We applied a principal-agent modeling approach to understanding how an insurance carrier can best manage its portfolio risk of cyber insurance policies, given interdependent risks across its policy holders. Chapters 2 and 3 presented a series of models that led to the analysis of a practical underwriting scenario in Chapter 4. In a nutshell, we demonstrated that simultaneously insuring risk-dependent agents can lead to higher profit, compared with not insuring them simultaneously, the reason being that the insurer can incentivize the source of the risk (e.g., a service provider or SP) to improve its security by offering a discount on its premium. When an SP provides more secure services for its customers, the chance of business interruption for the customers decreases, and the insurer's profit improves. In other words, receiving premiums from all interdependent agents and paying less in coverage due to improved security drives the profit opportunity not present when insuring interdependent agents.

In addition, we showed that even when the insurer, underwriting only the customers, is able to attribute a part of the loss to the SP and receive compensation from the latter's insurer due to third-party liability, its profit is still below that of insuring both the SP and its customers. The reason is that the insurer loses the SP's premium and the insurer cannot incentivize the SP to decrease the chance of business interruption for SP's customers. These results refute conventional wisdom that insurers should avoid insuring interdependent agents.

These results should help insurers and reinsurers better understand and manage systemic risk, while also demonstrating how market-based insurance can improve social welfare. There are a number of broad categories of open problems in research and practice.

Contingencies on periodic pre-screening: active policy.   Our models have assumed that the agent exerts a one-shot effort, which applies to the entire policy period. Under this assumption, pre-screening helps incentivize non-zero effort. In reality, keeping risk at a certain level typically requires sustained effort throughout the period, and it is conceivable that the insured may choose to lower his effort after the initial risk assessment (yet another form of moral hazard). If so, then our results on pre-screening suggests that it has to be performed more often, and repeated premium adjustment is made following each screening. This effectively means that the initial contract is an *active policy* with built-in contingencies, and the actual premium payable is realized over time depending on the screening results. We illustrate this idea using the following example with one additional, mid-term, screening.

Let's assume that the agent exerts effort e before the first screening, resulting in assessment outcome $S = e + W$ as before, and then he lowers the effort to e'. Accordingly, let $S' = e' + W'$ be the outcome of the second, mid-term screening, where W' is a zero-mean Gaussian noise with variance σ^2 and independent of W. Below we show that the insurer is able to incentivize the agent *not* to decrease the effort level through the second screening, i.e., to ensure $e' = e$.

Consider an active contract with three parameters (b, θ, θ') offered to the agent, where b is the base premium and θ a discount factor as before, and θ' is an additional, *penalty* factor. The total cost of the agent is given by:

$$X = b - \theta S + ce + \theta'(S - S') - c'(e - e'), \tag{6.1}$$

where $0 \leq c' \leq c$ is the benefit of lowering the effort and $\theta'(S - S')$ is the penalty that the insured pays after the second risk assessment; if $S - S' < 0$, then this term is effectively a reward for assessed increase in effort.

Similar to (2.21), the agent's expected utility under contract (b, θ, θ') is:

$$\overline{U}(b, \theta, e, \theta', e') = E(g(-X)) =$$
$$- \exp\{\gamma b + \gamma(c - c' + \theta' - \theta)e + \gamma(-\theta' + c')e' + \gamma^2 \sigma^2 \frac{(\theta - \theta')^2 + (\theta')^2}{2}\}. \tag{6.2}$$

The insurer's problem can be written as follows:

$$R(\sigma) = \max_{\{b, \theta, e, \theta', e'\}} E\left[b - \theta S + \theta'(S - S')\right] - p(e')L \tag{6.3}$$

$$\text{s.t.} \quad \text{(IR)} \quad \overline{U}(b, \theta, e, \theta', e') \geq u^o$$

$$\text{(IC)} \quad (e, e') \in \arg\max_{\tilde{e}, \tilde{e}'} \overline{U}(b, \theta, \tilde{e}, \theta', \tilde{e}'), \ e' \leq e.$$

The following theorem from [74] shows that the second risk assessment is effective in preventing the agent from lowering his effort.

Theorem 6.1   *Let $\hat{e}$ and $\hat{e}'$ be the agent's effort level at the solution to* (6.3), *and e^* be the optimal effort level in optimization problem* (2.24). *We have $\hat{e} = \hat{e}'$, and the optimal contract parameters are $\theta = c$ and $\theta' = c'$ if $\hat{e} > 0$; otherwise $\theta = \theta' = 0$. Moreover, if $e^* > 0$, then $\hat{e} = e^*$.*

Last, we have $V(\sigma) \leq R(\sigma)$, where $V(\sigma)$ is the insurer's maximum utility obtained from (2.24) *by assuming there is only one pre-screening and the agent does not lower his effort afterward, with equality achieved if $c' = c$.*

The last part of the above theorem suggests that performing the second screening helps the insurer to improve profit even when the agent may be assumed not to lower his effort. This is because the second pre-screening decreases the variance and uncertainty in the agent's utility. Therefore, a risk-averse agent is willing to pay more premium when the uncertainty and variance on his side decreases.

This is a fairly simplistic model with a single element of contingency, the outcome of a mid-term screening. It is not hard to imagine there could be other forms of contingencies built into the contract as additional incentive mechanisms, to further mitigate moral hazard.

Modeling aggregated and correlated risks.   This book is primarily focused on the risk of an adverse event, rather than the magnitude of that event. The latter is often referred to as Value-at-Risk (VaR) [98], and is a key piece of information in calculating premium. As we saw in Chapter 4, the rate-schedule based underwriting embeds the VaRs in the base premium and base retention values associated with a firm's size, revenue, and sector. Insurance practitioners often use Monte Carlo simulation to estimate the values of various loss scenarios, and this remains an active area of research.

There is ample evidence that cyber risks are heavy tailed [34, 44, 85], and there is generally a fear of black-swan events. The challenge of dealing with heavy-tailed cyber-risks is particularly prominent in market settings that manage aggregate cyber risks, which arise in systems with a larger number of components dependent on one another, e.g., manufacturing supply chain. This arises in major systemic cyber-attacks such as those caused by botnets and ransomware, resulting in correlated damages across organizations. Either scenario makes managing aggregate cyber-risk a very challenging proposition for a cyber (re)insurer simply because she might need to aggregate correlated heavy-tailed risks from multiple sources and the sum would be a risk that has a variance that increases with the number of risk sources. This could very well discourage a profit-seeking (and risk-averse) insurer from entering the market. A future research direction is to design insurance policies for such aggregate risks that enable insurers to profitably expand the coverage market in catastrophe cyber-risk scenarios, thereby providing loss support to clients needing insurance.

Closing the gap. Despite its potential and visible progress over the past decades, the cyber insurance services have been consistently lagging behind demand. The annual estimates put cyber loss coverage demand at $600 billion globally (1% of U.S. GDP) [39]. However, the total annual market for cyber insurance services is two orders of magnitude less (estimated to be $6 billion globally [124]). Such a low market penetration has been attributed to reasons including a lack of coverage awareness on the demand side (prospective insureds), a perceived mismatch between what firms expect an insurer would cover versus what coverages are contractually provided, human tendency to underestimate bad risk and overestimate good risks [116], and a lack of quality data that contributed to pricing nuances [100, 124], and fear of substantial correlated and aggregated risk on the supply side as mentioned earlier, especially when they are not equipped with measures to robustly estimate and alleviate cyber risks. These factors point to open challenges that need to be addressed to grow the cyber insurance market for the social good.

Quantifying the social cost of data breaches. Value-at-risk is singularly focused on the direct cost to the entity at risk. However, data breaches carries substantial indirect, social costs. We have argued earlier on that cybersecurity (or investment in cybersecurity) should be viewed as public good. Within this context, there is a parallel between carbon emission and (poor) cybersecurity in terms of their (negative) externalities. The negative externalities linked to carbon emission are becoming increasingly clear [54]: adverse impact on climate change due to carbon emission is a form of social costs imposed on all, including those who did not participate or benefit from the carbon-related economic activity. Similarly, the negative externality of one firm's poor effort in governing its cybersecurity has been discussed at length in earlier chapters in the context of vendor and supply chain relationships.

The difference between the two is that there is now substantial research into estimating the indirect, social cost of carbon, e.g., it has been estimated at roughly $50/metric ton of CO_2 generated [28, 54]. By contrast, estimating the social cost of data breaches is still in its infancy; the public awareness of this social cost is similarly lagging. There is plenty of studies on estimating the direct cost associated with a data breach or cyber incident, such as forensics, crisis management, legal fees, public relations, business interruption to their immediate customers, system replacement, etc. However, it is not at all clear what the costs are borne by the public whose records were stolen due to the breach and the (criminal) activities that follow: individuals can fall victim to not just identity theft (that can take years and thousands of dollars to clean up), but also blackmail and extortion, and whatever emotional distress and medical conditions they induce; breaches such as that happened at Ashley Madison have led to public shaming, extortion, hate crime, and even suicide [1]; stolen healthcare records enable insurance fraud whose cost initially borne by the insurance companies is also invariably transferred to the premium-paying public.

These are important questions that need to be addressed both to inform policy making and to educate the public.

Bibliography

[1] Ashley Madison Data Breach. https://en.wikipedia.org/wiki/Ashley_Madison_data_breach 116

[2] Benchmarking Trends: Cyber-Attacks Drive Insurance Purchases For New and Existing Buyers. https://www.marsh.com/content/dam/marsh/Documents/PDF/US-en/Mid-Year%20Cyber%20Benchmarking%20Report-10-15.pdf 1

[3] Betterley Reports. https://www.irmi.com/free-resources/authoritative-reports/betterley-reports 1

[4] Blacklist (computing). https://en.wikipedia.org/wiki/Blacklist_(computing) 98

[5] Composite Blocking List. http://cbl.abuseat.org/ 98

[6] Cyber Security Insurance. https://taigusa.com/cybersecurity-insurance/ 1

[7] *Cyber Security Professional Trends: A SANS Survey.* http://bit.ly/2rulxon 85

[8] Darknet. https://en.wikipedia.org/wiki/Darknet 98

[9] Deep Packet Inspection. https://en.wikipedia.org/wiki/Deep_packet_inspection 109

[10] DShield. http://www.dshield.org/ 99

[11] Environmental, Social and Corporate Governance. https://en.wikipedia.org/wiki/Environmental,_social_and_corporate_governance 110

[12] hpHosts for your Protection. http://hosts-file.net/ 98

[13] Multiple DNS Implementations Vulnerable to Cache Poisoning. http://www.kb.cert.org/vuls/id/800113 97

[14] NetDilligence Cyber Claims Study. https://netdiligence.com/cyber-claims-study-2019-report/ 1

[15] Neustar IP Intelligence Data. 97, 99, 106

[16] Open Resolver Project. http://openresolverproject.org/ 97

[17] OpenBL. http://www.openbl.org/ 99

[18] Peering. https://en.wikipedia.org/wiki/Peering 109

[19] PhishTank. http://www.phishtank.com/ 98

[20] SpamCop Blocking List. http://www.spamcop.net/ 98

[21] Spamtrap. https://en.wikipedia.org/wiki/Spamtrap 98

[22] Supervised Learning. https://en.wikipedia.org/wiki/Supervised_learning 101

[23] SURBL: Reputation Data. http://www.surbl.org/ 98

[24] The SPAMHAUS Project: SBL, XBL, PBL, ZEN Lists. http://www.spamhaus.org/ 98

[25] U-M Disinvests in Fossil Fuel as Part of Goal to have Net Zero Carbon Investment Portfolio. https://www.freep.com/story/news/education/2021/03/25/university-of-michigan-disinvests-fossil-fuel/7000151002/ 111

[26] UCEPROTECTOR Network. http://www.uceprotect.net/ 98

[27] WPBL: Weighted Private Block List. http://www.wpbl.info/ 98

[28] The True Cost of Carbon Pollution. https://www.edf.org/true-cost-carbon-pollution 2021. 116

[29] Alexa Web Information Service. http://aws.amazon.com/awis 106

[30] R. Anderson. Why information security is hard—an economic perspective. In *17th Annual Computer Security Applications Conference*, pages 358–365, 2001. DOI: 10.1109/acsac.2001.991552 9

[31] R. Anderson and T. Moore. Information security economics and beyond. In *Annual International Cryptology Conference*, pages 68–91, 2007. DOI: 10.1007/978-3-540-74143-5_5 10

[32] T. P. Augustinos, L. Bauer, A. Cappelletti, J. Chaudhery, I. Goddijn, L. Heslault, N. Kalfigkopoulos, V. Katos, N. Kitching, M. Krotofil, and E. Leverett. Cyber-insurance: Recent advances, good practices and challenges. In *European Union Agency for Network and Information Security (ENISA)*, 2016. 10

[33] Betterley. The betterley report: Cyber/privacy insurance market survey, June 2014. http://betterley.com/samples/cpims14_nt.pdf 2

[34] C. Biener, M. Eling, and J. Wirfs. Insurability of cyber risk: An empirical analysis. *The Geneva Papers on Risk and Insurance—Issues and Practice*, 40(1):131–158, 2015. DOI: 10.1057/gpp.2014.19 115

[35] R. Böhme and G. Kataria. Models and measures for correlation in cyber-insurance. In *The Workshop on the Economics of Information Security (WEIS)*, 2006. 11

[36] R. Böhme and G. Schwartz. Modeling cyber-insurance: Towards a unifying framework. In *The Workshop on the Economics of Information Security*, 2010. 10

[37] J. Bolot and M. Lelarge. Cyber-insurance as an incentive for internet security. In *The Workshop on the Economics of Information Security*, 2008. DOI: 10.1007/978-0-387-09762-6_13 11

[38] P. Bolton and M. Dewatripont. *Contract Theory*. MIT Press, 2005. 2

[39] A. Coburn, E. Leverett, and G. Woo. *Solving Cyber Risk: Protecting Your Company and Society*. John Wiley & Sons, 2018. 1, 116

[40] The Web Application Security Consortium. Web-Hacking-Incident-Database. http://projects.webappsec.org/w/page/13246995/Web-Hacking-Incident-Database 96

[41] J. Damas and F. Neves. Preventing use of recursive nameservers in reflector attacks, 2008. https://tools.ietf.org/html/rfc5358 97

[42] Z. Durumeric, E. Wustrow, and J. A. Halderman. ZMap: Fast internet-wide scanning and its security applications. In *Proc. of the 22nd USENIX Security Symposium*, pages 605–620, Washington, DC, August 2013. 97

[43] K. M. Eisenhardt. Agency theory: An assessment and review. *The Academy of Management Review*, 14(1):57–74, 1989. DOI: 10.5465/amr.1989.4279003 2

[44] M. Eling and J. Wirfs. What are the actual costs of cyber risk events? *European Journal of Operational Research*, 272(3):1109–1119, 2019. DOI: 10.1016/j.ejor.2018.07.021 115

[45] Risk Engineering. Insurance and Risk: Some History. https://risk-engineering.org/concept/history-of-insurance xiii, 1

[46] S. Evelsizer and B. Eaton. Netdiligence 2016 cyber claims study, 2016. https://netdiligence.com/wp-content/uploads/2016/10/P02_NetDiligence-2016-Cyber-Claims-Study-ONLINE.pdf 1, 74, 78

[47] M. Ezhei and B. T. Ladani. Interdependency analysis in security investment against strategic attacks. *Information Systems Frontiers*, April 2018. DOI: 10.1007/s10796-018-9845-8 37

[48] F. Farhadi, H. Tavafoghi, D. Teneketzis, and J. Golestani. A dynamic incentive mechanism for security in networks of interdependent agents. In *Game Theory for Networks: 7th International EAI Conference GameNets, Proceedings*, pages 86–96, Springer International

Publishing, Cham, Knoxville, TN, May 9, 2017. DOI: 10.1007/978-3-319-67540-4_8 37

[49] World Economic Forum. Understanding systemic cyber risk. In *Global Agenda Council on Risk and Resilience*, October 2016. 37

[50] D. Fudenberg and J. Tirole. *Game Theory*. MIT Press, 1993. 2

[51] Gartner. Security technology market to reach $86B in 2016, June 2013. http://www.eweek.com/small-business/security-technology-market-to-reach-86b-in-2016-gartner 110

[52] Gartner. Gartner forecasts worldwide security and risk management spending growth to slow but remain positive in 2020, June 2020. https://www.gartner.com/en/newsroom/press-releases/2020-06-17-gartner-forecasts-worldwide-security-and-risk-managem 110

[53] R. Gibbons. *A Primer in Game Theory*. Harvester-Wheatsheaf, 1992. 2

[54] J. T. Gordon. How accounting for the externalities of carbon would affect end products, December 2019. https://www.atlanticcouncil.org/blogs/energysource/how-accounting-for-the-externalities-of-carbon-would-affect-end-products/ 116

[55] J. Grossklags, S. Radosavac, A. A. Cárdenas, and J. Chuang. Nudge: Intermediaries' role in interdependent network security. In *Trust and Trustworthy Computing*, pages 323–336, Springer, 2010. DOI: 10.1007/978-3-642-13869-0_24 9

[56] S. A. Hasheminasab and B. T. Ladani. Security investment in contagious networks. *Risk Analysis*, 2018. DOI: 10.1111/risa.12966 37

[57] C. Hemenwa. ABI Research: Cyber-insurance market to reach $10B by 2020, 2015. www.advisenltd.com/2015/07/30/abi-research-cyber-insurance-market-to-reach-10b-by-2020/, 1

[58] D. T. Hoang, D. Niyato, and P. Wang. Optimal cost-based cyber insurance policy management for mobile services. In *IEEE 86th Vehicular Technology Conference (VTC-Fall)*, pages 1–5, September 2017. DOI: 10.1109/vtcfall.2017.8288224 6

[59] A. Hofmann. Internalizing externalities of loss prevention through insurance monopoly: An analysis of interdependent risks. *The Geneva Risk and Insurance Review*, 2007. DOI: 10.1007/s10713-007-0004-2 10, 11

[60] A. Hubert and R. Van Mook. Measures for making DNS more resilient against forged answers. RFC 5452, January 2009. DOI: 10.17487/rfc5452 97

[61] Insurance Information Institute.   U.S. cyber insurance market demonstrates growth, innovation in wake of high profile data breaches, 2015.   www.iii.org/pressrelease/us-cyber-insurance-market-demonstrates-growthinnovation-in-wake-of-high-profile-data-breaches-102015 1

[62] L. Jiang, V. Anantharam, and J. Walrand.   How bad are selfish investments in network security? *IEEE/ACM Transactions on Networking*, 19(2):549–560, April 2011. DOI: 10.1109/tnet.2010.2071397 16

[63] L. Jiang, V. Anantharam, and J. Walrand.   How bad are selfish investments in network security? *IEEE/ACM Transactions on Networking*, 19(2):549–560, April 2011. DOI: 10.1109/tnet.2010.2071397 71

[64] B. Johnson, J. Grossklags, N. Christin, and J. Chuang.   Uncertainty in interdependent security games.   In *International Conference on Decision and Game Theory for Security*, pages 234–244, Springer, 2010. DOI: 10.1007/978-3-642-17197-0_16 37

[65] B. Johnson, J. Grossklags, N. Christin, and J. Chuang.   Are security experts useful? Bayesian nash equilibria for network security games with limited information.   In *European Symposium on Research in Computer Security*, pages 588–606, Springer, 2010. DOI: 10.1007/978-3-642-15497-3_36 37

[66] J. Kesan, R. Majuca, and W. Yurcik.   Cyber-insurance as a market-based solution to the problem of cyber security: A case study. In *Workshop on the Economics of Information Security (WEIS)*, pages 1–46, 2005. 9

[67] J. Kesan, R. Majuca, and W. Yurcik.   The economic case for cyber-insurance. University of Illinois Legal Working Paper Series UIUCLWPS-1001, University of Illinois College of Law, 2004. 10, 11

[68] J. P. Kesan, R. P. Majuca, and W. Yurcik.   Cyber-insurance as a market-based solution to the problem of cyber security. In *The Workshop on the Economics of Information Security*, 2005. 10, 11

[69] M. Khalili, M. Liu, and S. Romanosky.   Embracing and controlling risk dependency in cyber-insurance policy underwriting.   In *Annual Workshop on the Economics of Information Security (WEIS)*, Innsbruck, Austria, June 2018.   Journal version appearing in *Journal of Cyber Security*, accepted for publication. DOI: 10.1093/cybsec/tyz010 9, 11

[70] M. Khalili, M. Liu, and S. Romanosky.   Embracing and controlling risk dependency in cyber-insurance policy underwriting. *Journal of Cyber Security*, 5(1), October 2019. DOI: 10.1093/cybsec/tyz010 9, 11, 64, 69, 78

[71] M. Khalili, P. Naghizadeh, and M. Liu. Designing cyber-insurance policies in the presence of security interdependence. In *The 12th Workshop on the Economics of Networks, Systems and Computation (NetEcon)*, Boston, MA, June 2017. DOI: 10.1145/3106723.3106730 9, 11, 46

[72] M. Khalili, P. Naghizadeh, and M. Liu. Embracing risk dependency in designing cyber-insurance contracts. In *Annual Allerton Conference on Control, Communication, and Computing (Allerton)*, Allerton, IL, October 2017. DOI: 10.1109/allerton.2017.8262837 9, 11

[73] M. Khalili, P. Naghizadeh, and M. Liu. Designing cyber-insurance policies: The role of pre-screening and security interdependence. *IEEE Transactions on Information Forensics and Security (TIFS)*, 13(9):2226–2239, September 2018. DOI: 10.1109/tifs.2018.2812205 9, 11, 46, 47

[74] M. Khalili, X. Zhang, and M. Liu. Effective premium discrimination for designing cyber-insurance policies with rare losses. In *Conference on Decision and Game Theory for Security (GameSec)*, Stockholm, Sweden, October 2019. DOI: 10.1007/978-3-030-32430-8_16 27, 115

[75] H. Kunreuther and G. Heal. Interdependent security. *Journal of Risk and Uncertainty*, 26(2–3):231–249, 2003. DOI: 10.1023/A:1024119208153 9, 37

[76] R. J. La. Interdependent security with strategic agents and cascades of infection. *IEEE/ACM Transactions on Networking*, 24(3):1378–1391, June 2016. DOI: 10.1109/tnet.2015.2408598 38

[77] R. J. La. Effects of degree correlations in interdependent security: Good or bad? *IEEE/ACM Transactions on Networking*, 25(4):2484–2497, August 2017. DOI: 10.1109/tnet.2017.2691605 37

[78] J. J. Laffont and D. Martimort. *The Theory of Incentives: The Principal-Agent Model*. Princeton University Press, 2009. 2

[79] A. Laszka, M. Felegyhazi, and L. Buttyan. A survey of interdependent information security games. *ACM Computing Surveys*, 47(2):23:1–23:38, August 2014. DOI: 10.1145/2635673 16

[80] A. Laszka and G. Schwartz. *Becoming Cybercriminals: Incentives in Networks with Interdependent Security*, pages 349–369. Springer International Publishing, Cham, 2016. DOI: 10.1007/978-3-319-47413-7_20 37

[81] M. Lelarge and J. Bolot. Economic incentives to increase security in the internet: The case for insurance. In *IEEE International Conference on Computer Communications (INFOCOM)*, 2009. DOI: 10.1109/infcom.2009.5062066 10, 11

[82] M. Lelarge. Coordination in network security games: A monotone comparative statics approach. *IEEE Journal on Selected Areas in Communications*, 30(11):2210–2219, 2012. DOI: 10.1109/jsac.2012.121213 38, 71

[83] A. Liaw and M. Wiener. Classification and Regression by Random Forest, 2002. http://CRAN.R-project.org/doc/Rnews/ 103

[84] Y. Liu, A. Sarabi, J. Zhang, P. Naghizadeh, M. Karir, M. Bailey, and M. Liu. Cloudy with a chance of breach: Forecasting cyber security incidents. In *USENIX Security Symposium*, Washington, DC, August 2015. 9, 95, 101

[85] T. Maillart and D. Sornette. Heavy-tailed distribution of cyber-risks. *The European Physical Journal B*, 75(3):357–364, 2010. DOI: 10.1140/epjb/e2010-00120-8 115

[86] Marsh. Benchmarking trends: More companies purchasing cyber-insurance, March 2013. https://www.marsh.com/us/insights/research/benchmarking-trends-more-companies-purchasing-cyber-insurance.html 2

[87] A. Mas-Colell, M. D. Whinston, and J. R. Green. *Microeconomic Theory*. Oxford University Press, New York, 1995. 9, 11

[88] F. Massacci, J. Swierzbinski, and J. Williams. Cyber-insurance and public policy: Self-protection and insurance with endogenous adversaries, 2017. 71

[89] R. A. Miura-Ko, B. Yolken, N. Bambos, and J. Mitchell. Security investment games of interdependent organizations. In *Proc. of 46th Annual Allerton Conference on Communication, Control, and Computing*, pages 252–260, 2008. DOI: 10.1109/allerton.2008.4797564 37

[90] H. Ogut, N. Menon, and S. Raghunathan. Cyber-insurance and it security investment: Impact of interdependence risk. In *The Workshop on the Economics of Information Security*, 2005. 10, 11

[91] M. J. Osborne and A. Rubenstein. *A Course in Game Theory*. MIT Press, 1994. 2

[92] R. Pal and L. Golubchik. Analyzing self-defense investments in internet security under cyber-insurance coverage. In *IEEE 30th International Conference on Distributed Computing Systems*, pages 339–347, 2010. DOI: 10.1109/icdcs.2010.79 10, 11

[93] R. Pal, L. Golubchik, K. Psounis, and P. Hui. On a way to improve cyber-insurer profits when a security vendor becomes the cyber-insurer. In *IFIP Networking Conference*, pages 1–9, May 2013. 11

[94] R. Pal, L. Golubchik, K. Psounis, and P. Hui. Will cyber-insurance improve network security? A market analysis. In *IEEE Conference on Computer Communications (INFOCOM)*, pages 235–243, 2014. DOI: 10.1109/infocom.2014.6847944 10, 11

[95] R. Pal, L. Golubchik, K. Psounis, and P. Hui. Will cyber-insurance improve network security? A market analysis. In *IEEE International Conference on Computer Communications (INFOCOM)*, 2014. DOI: 10.1109/infocom.2014.6847944 11, 22

[96] R. Pal, L. Golubchik, K. Psounis, and P. Hui. Security pricing as enabler of cyber-insurance a first look at differentiated pricing markets. *IEEE Transactions on Dependable and Secure Computing*, PP(99):1, 2017. 10, 71

[97] P. Passeri. http://hackmageddon.com/ 96

[98] R. T. Rockafellar and S. Uryasev. Conditional value-at-risk for general loss distributions. *Journal of Banding and Finance*, 26:1443–1471, 2002. DOI: 10.1016/s0378-4266(02)00271-6 8, 115

[99] S. Romanosky. Comments to the department of commerce on incentives to adopt improved cyber security practices, April 2013. https://www.ntia.doc.gov/files/ntia/romanosky_comments.pdf 2

[100] S. Romanosky, L. Ablon, A. Kuehn, and T. Jones. Content analysis of cyber-insurance policies: How do carriers price cyber-risk? *Journal of Cyber Security*, 5(1), 2019. 116

[101] S. Romanosky, L. Ablon, A. Kuehn, and T. Jones. Content analysis of cyber-insurance policies: How do carriers write policies and price cyber-risk? *Journal of Cyber Security*, 5(1), February 2019. 6, 10, 67

[102] A. Rubinstein and M. E. Yaari. Repeated insurance contracts and moral hazard. *Journal of Economic Theory*, 30(1):74–97, 1983. DOI: 10.1016/0022-0531(83)90094-7 27

[103] A. Sarabi, P. Naghizadeh, Y. Liu, and M. Liu. Prioritizing security spending: A quantitative analysis of risk distributions for different business profiles. In *Annual Workshop on the Economics of Information Security (WEIS)*, Delft, The Netherlands, June 2015. 9, 95, 105

[104] A. Sarabi, P. Naghizadeh, Y. Liu, and M. Liu. Risky business: Fine-grained data breach prediction using business profiles. *Journal of Cyber Security*, 2(1):15–28, December 2016. DOI: 10.1093/cybsec/tyw004 9

[105] G. Schwartz, N. Shetty, and J. Walrand. Cyber-insurance: Missing market driven by user heterogeneity, 2010. www.eecs.berkeley.edu/?nikhils/sectypes.pdf 10

[106] G. A. Schwartz and S. S. Sastry. Cyber-insurance framework for large scale interdependent networks. In *The 3rd International Conference on High Confidence Networked Systems*, 2014. DOI: 10.1145/2566468.2566481 10, 11

[107] N. Shetty, G. Schwartz, M. Felegyhazi, and J. Walrand. Competitive cyber-insurance and internet security. *Economics of Information Security and Privacy*, pages 229–247, 2010. DOI: 10.1007/978-1-4419-6967-5_12 10, 11, 22

[108] N. Shetty, G. Schwartz, and J. Walrand. Can competitive insurers improve network security? In *The 3rd International Conference on Trust and Trustworthy Computing (TRUST)*, 2010. DOI: 10.1007/978-3-642-13869-0_23 10

[109] S. Shetty, M. McShane, L. Zhang, J. P. Kesan, C. A. Kamhoua, K. Kwiat, and L. L. Njilla. Reducing informational disadvantages to improve cyber-risk management. *The Geneva Papers on Risk and Insurance—Issues and Practice*, February 2018. DOI: 10.1057/s41288-018-0078-3 37

[110] A. Shortland. *Kidnap: Inside the Ransom Business*. Oxford University Press, April 2019. DOI: 10.1093/oso/9780198815471.001.0001 6

[111] M. Simaan and Jr. J. B. Cruz. On the Stackelberg strategy in nonzero-sum games. *Journal of Optimization Theory and Applications*, 11(5):533–555, May 1973. DOI: 10.1007/bf00935665 2

[112] B. Stone-Gross, M. Cova, L. Cavallaro, B. Gilbert, M. Szydlowski, R. Kemmerer, C. Kruegel, and G. Vigna. Your botnet is my botnet: Analysis of a botnet takeover. In *Proc. of CCS*, 2009. DOI: 10.1145/1653662.1653738 98

[113] J. Strickland. 10 worst computer viruses of all time. https://computer.howstuffworks.com/worst-computer-viruses.htm/printable/ 38

[114] O. Thonnard, L. Bilge, A. Kashyap, and M. Lee. Are you at risk? Profiling organizations and individuals subject to targeted attacks. In *Financial Cryptography and Data Security*, January 2015. DOI: 10.1007/978-3-662-47854-7_2 106

[115] D. K. Tosh, I. Vakilinia, S. Shetty, S. Sengupta, C. A. Kamhoua, L. Njilla, and K. Kwiat. Three layer game theoretic decision framework for cyber-investment and cyber-insurance. In Stefan Rass, Bo An, Christopher Kiekintveld, Fei Fang, and Stefan Schauer, Eds., *Decision and Game Theory for Security*, pages 519–532, Springer International Publishing, Cham, 2017. DOI: 10.1007/978-3-319-68711-7_28 6

[116] A. Tversky and D. Kahneman. Advances in prospect theory: Cumulative representation of uncertainty. *Journal of Risk and Uncertainty*, 5(4):297–323, 1992. DOI: 10.1007/bf00122574 116

[117] I. Vakilinia, S. Badsha, and S. Sengupta. Crowdfunding the insurance of a cyber-product using blockchain. In *Ubiquitous Computing, Electronics and Mobile Communication Conference (UEMCON)*, 2018. DOI: 10.1109/uemcon.2018.8796515 11

[118] I. Vakilinia and S. Sengupta. A coalitional cyber-insurance framework for a common platform. *IEEE Transactions on Information Forensics and Security*, 2018. DOI: 10.1109/tifs.2018.2881694 11

[119] H. Varian. System reliability and free riding. *Economics of Information Security*, pages 1–15, 2004. DOI: 10.1007/1-4020-8090-5_1 9

[120] VERIS Community Database (VCDB). http://vcdb.org 96

[121] Verizon. Data Breach Investigations Reports (DBIR) 2014. http://www.verizonenterprise.com/DBIR/ 96, 107, 108

[122] Verizon. Data Breach Investigations Report (DBIR) 2020. https://enterprise.verizon.com/resources/reports/dbir/ 110

[123] H. von Stackelberg. *Market Structure and Equilibrium*, 1st ed., Springer, 2011. 2

[124] S. S. Wang. Integrated framework for information security investment and cyber-insurance. *Pacific-Basin Finance Journal*, 57(101173), 2019. DOI: 10.1016/j.pacfin.2019.101173 116

[125] Z. Yang and J. C. S. Lui. Security adoption and influence of cyber-insurance markets in heterogeneous networks. *Performance Evaluation*, 74:1–17, April 2014. DOI: 10.1016/j.peva.2013.10.003 10, 11

[126] J. Zhang, Z. Durumeric, M. Bailey, M. Liu, and M. Karir. On the mismanagement and maliciousness of networks. In *Network and Distributed System Security Symposium (NDSS)*, San Diego, CA, February 2014. DOI: 10.14722/ndss.2014.23057 95

[127] X. Zhao, L. Xue, and A. B. Whinston. Managing interdependent information security risks: Cyber-insurance, managed security services, and risk pooling arrangements. *Journal of Management Information Systems*, 30(1):123–152, 2013. DOI: 10.2753/mis0742-1222300104 38

Author's Biography

MINGYAN LIU

Mingyan Liu is the Peter and Evelyn Fuss Chair of Electrical and Computer Engineering at the University of Michigan. She received her B.Sc. in electrical engineering in 1995 from Nanjing University of Aeronautics and Astronautics, Nanjing, China, M.Sc. in systems engineering, and Ph.D. in electrical engineering from the University of Maryland, College Park, in 1997 and 2000, respectively. She joined the Department of Electrical Engineering and Computer Science at the University of Michigan, Ann Arbor, in September 2000, where she is currently a Professor.

Her research interests are in optimal resource allocation, sequential decision theory, incentive design, and performance modeling and analysis, all within the context of large-scale networked systems. Her most recent research activities involve cyber risk quantification and designing cybersecurity incentive mechanisms using large-scale Internet measurement data and machine learning techniques. She is the recipient of the 2002 NSF CAREER Award, the University of Michigan Elizabeth C. Crosby Research Award in 2003 and 2014, the 2010 EECS Department Outstanding Achievement Award, the 2015 College of Engineering Excellence in Education Award, the 2017 College of Engineering Excellence in Service Award, and the 2018 Distinguished University Innovator Award. She is a Fellow of the IEEE and a member of the ACM.

Printed in the United States
by Baker & Taylor Publisher Services